WHEN EVOLUTION STOPS

A Nexus of Geology, Astronomy and Biology

Armando Simón

© 2004, 2007, 2017, 2018, 2023

Published by Lulu Publishers

Raleigh, North Carolina

ISBN: 978-1-304-85744-6

The idea is like truth itself, so simple and obvious that those who read and understand it will be struck by its simplicity; yet it is perfectly original.
      ---Henry Bates

My reflection when I first made myself master of the central idea was, "How extremely stupid not to have thought of that."
      ----Thomas Henry Huxley

Like many great scientific ides, from Newton's theory of gravitation to Darwin's theory of natural selection, Nash's idea seemed initially too simple to be truly interesting, too narrow to be widely applicable, and, later on, so obvious that its discovery by *someone* was deemed all but inevitable.
---Sylvia Nasar, *A Beautiful Mind*

As so often happens when the obvious becomes obvious, no one could understand how in the world "the obvious" had not been recognized earlier for what it was.
---Jeffrey Schwartz, *Sudden Origins*

Sometimes being an interaction designer is so frustrating! If, as a designer you do something really, fundamentally, blockbuster correct, everybody looks at it and says, "Of course! What *other* way would there be?" This is true even if the client has been staring, empty-handed and idea-free, at the problem for months or even years without a clue about solving the problem. It's also true even if our solution generates millions of dollars for the company. Most really breakthrough conceptual advances are *opaque in foresight and transparent in hindsight.* It is incredibly hard to see breakthroughs in design. You can be trained and prepared, spend hours studying the problem, and still not see the answer. Then someone else comes along and points out a key insight, and the vision clicks into place with the natural obviousness of the wheel. If you shout the solution from the rooftops, others will say, "Of course the wheel is round! What other shape could it possibly be?" This makes it frustratingly hard to show off good design work.

---Alan Cooper, *The Inmates Are Running the Asylum*

And then, of course, success once achieved and disseminated is redescribed as common sense, and we are assured that any sensible person can see there is only one way---the right way.

---David Wooton. *The Invention of Science*

TABLE OF CONTENTS

Introduction

Chapter 1: The First Mechanism: Spirits

Chapter 2: The Second Mechanism: Spontaneous Generation

Chapter 3: The Third Mechanism: Lamarck

Chapter 4: The Fourth Mechanism: Natural Selection

Chapter 5: The Fifth Mechanism: Mutations

Chapter 6: Back to the First Mechanism

Chapter 7: The Sixth Mechanism: Symbiogenesis

Chapter 8: The Flaws in the Classical Theory

Chapter 9: The McClintock Effect: The Seventh Mechanism

Conclusion

References

Appendix

A RATHER LENGTHY INTRODUCTION

Our educational system is full of subdivisions that are artificial, that shouldn't be there.
---Barbara McClintock

[I] happen to be by birth . . . absolutely impervious to authority unable to demonstrate its tenets on grounds other than authority. ---Leon Croizat

Scientists often---perhaps usually---find it hard to let go of a theory that they care about. When some devastating new finding shows it to be wrong, that's hard for them to accept. There are many theories whose proponents have clung onto them for too long, rending them more and more elaborate in a desperate attempt to accommodate the findings that disprove them and ward off the inevitable end.
---Gabrielle Walker, *Snowball Earth*

I began this work with great trepidation. I do not consider myself to be a timid person, and no one has ever accused me of that---quite the contrary---yet, I have to admit that I was very, very, hesitant to plunge in. It took me a surprisingly long time to write this small treatise; also, in my handwritten draft, I was surprised at how choppy and disorganized my writing was, whereas, usually, whenever I write a paper it is an evenly smooth flow of narration.

Being originally from Cuba and growing up in a

very argumentative family, I was used to arguments and shouting matches; I often joke that we lived to argue. Once, when I was ten, I had the dubious honor of almost sending my family to a Communist prison when I got up in front of the class and poured forth an anti-Communist essay I had written. The horrified teacher, instead of right away reporting me to the secret police, called my parents, took them aside, and told my equally panic-stricken parents what had transpired. And, in the Sixties and Seventies in the United States, I held views, which were anti-totalitarian and, thusly, went strongly against the *zeitgeist* in universities. Lastly, in 1988 I had the really dubious honor of being (simultaneously!) fired from two universities, Valencia Community College and Seminole Community College, in Orlando, Florida, where I was a typically exploited adjunct professor in both places, for having written an unpublished essay, which I carelessly left lying around, satirizing feminists.

    The reasons for not writing down my thoughts on the topic of evolution were many. I am neither a biologist nor a paleontologist (though I had always wanted to be the latter, but my childish ambitions were derailed by the Cuban Diaspora, not that Cuba had schools specializing in that field), but a research psychologist. Notwithstanding perennial calls for generalists in science (Zirkle, 1959; Rose, 1976; Garwin, 1995; Ehrenfeld, 1995; McKinney, 1996; Margulis, 1998; Miele, 1998; Raup, 1999; Rugg & D'Agnese, 2013), the reality is harsher. Within each scientific field, the subspecialties have become so specialized that not only are journals dedicated to just those specific subspecialties and will not even consider submissions on topics that the editor feels may be inappropriate, but scientists in those fields, but belonging to other subspecialties, (Welch, 2017) sometimes have difficulty in understanding some of the articles in the specialty journals (Lynn Margulis referred to this state of affairs as "academic apartheid"). To be sure, there are many general journals that publish papers from all subspecialties within that particular scientific discipline. And, true, there are the two superb journals (*Science* in America and *Nature* in Great Britain) of world renown that deal with topics from every scientific field (though *Nature* refuses to publish psychological papers), but these two are the exception. Therefore, human nature being what it is, it would be only natural that the inevitable criticism of my position would emphasize, through snide remarks, that I was stepping out of my field of expertise, an ignorant interloper. And, of course, there would be some truth to that. I was particularly nervous about making some elementary mistake of facts in dealing in another discipline, not such an unlikely possibility[1] when one considers, as I said, that scientific fields are subdivided into specialties of which members of other specialties within the same field are ignorant about the details of other specialties. I took some comfort in the fact that in the history of science, contributions in one scientific field have occasionally been made by scientists from other fields: Gregor Mendel was a monk who laid the foundation for the science of genetics; Max Delbruck was a theoretical physicist who became one of the major proponents of molecular biology; Leo Szilard was another physicist who worked on the atomic bomb project and later did important work in biology (Keller, 1983); Heinrich Schwabe was a pharmacist who discovered that sunspots occur in cycles; William Herschel was a musician (Jones, 1988) who discovered the planet Uranus------which he at first thought was a comet; Francis Crick was another physicist who turned his attention to biology; Alfred Wegener was a meteorologist who innovated the concept of continental drift, John Raven was a classicist who uncovered a botanical hoax (Sabbagh, 1999), Jacobus van't Hoff worked out the scheme for the tetravalent carbon atom while teaching at a veterinary school (Goldsmith, 1977), Louis Pasteur was a chemist who was looked down by physicians as he elucidated the role of germs in various putrefactions; Luis Alvarez was a physicist when he postulated that a meteor impact wiped out the dinosaurs[2] and, by coincidence, just as I began on this project, a psychologist found an Inca city hidden in the Andes. So, you see, there is precedence.[3]

A second reason for hesitation on my part was my own uncertainty that *surely there must be more to it than that* kind of feeling towards the theory of evolution. But, no, it *was* that simple. One of the beauties of the theory is that it *is* so breathtakingly simple! In an era where the scientific disciplines have become so complex, so jargon laden, that its specialties can be incomprehensible to others the realization of its simplicity was to me a shock, a throwback to the early days of the naturalists where any intelligent person could pick up a scientific journal and follow the argument with little difficulty. And because it involves no math, anyone can understand it (although it is surprising how few actually do). This also explains why so many laymen have seen fit to comment on the theory.

Another, third, reason for showing the white feather was due to my upbringing. My father was a cardiologist who, whatever his faults, instilled in me a profound respect and awe for the achievements of the great men and women in science and the arts (a feeling that I have tried to pass on to my children in what has turned out to be a dismal exercise in futility). And I could not help but think, again and again, *Who am I, a third-rate psychologist, to challenge the conclusions of a giant like Charles Darwin?* That truly filled me with trepidation, that I would actually have the nerve to try to do so. I genuinely admired Darwin and Wallace. And I still do. Very much so. I always feel humble when reading Wallace or Darwin, or about them. Additionally, I am pained when I read caustic criticisms of the original theory, referred to as "Darwinism," because a lot of the sneering and anger by critics is actually directed, albeit indirectly and well deserved, at dogmatic neo-Darwinists; I am even upset at calling them neo-*Darwinists*. I can think of many other terms to call them (none of them publishable). William Bateson mirrored my outlook:

> It is more than thirty years since the *Origin of the Species* was written, but for many these questions are in no sense answered yet. In owning that it is so, we shall not honour Darwin's memory the less; for whatever may be the part which shall be finally assigned to Natural selection, it will always be remembered that it was through Darwin's work that men saw for the first time that the problem is one which man may reasonably hope to solve. If Darwin did not solve the problem himself, he first gave us the hope of a solution, perhaps a greater thing. (quote found in

Schwartz, 1999; p.196)

Hence, my criticism herein of the theory shall be referred to as the classical theory.

And, lastly, there was the question of mechanisms. It was not enough to point out that the theory, as it now stood, was deeply flawed. I felt that I had to offer a reasonable alternative---and I had none. Nonetheless, I naively proceeded with the intention of simply writing a critique. I had only begun to compile my notes in the summer of '03 when, in search of light reading, I happened to come across the writings of Barbara McClintock; suddenly, everything fell into place, everything made sense. I was so ebullient I had to get up and could not sit back down for hours. I felt like dancing---literally had the urge to dance (like I said, I am Cuban).

And, yet, reading her work, how did no one ever make the so obvious connection? How could someone not see the obvious evolutionary implications of her research? Indeed, in many of my other observations, put forth in the following chapters, the same thought has repeatedly crossed my mind: *How could they not see something so obvious?* Something that did not require knowledge of an esoteric subject, or a field expedition? I think it is due to what, in Psychology, we call a "mindset" and in common parlance, "not seeing the forest for the trees". To quote David Raup (1999): "But people outside a discipline can sometimes be very insightful, in part because they are not steeped in conventional ways of interpreting data in that discipline." (p. 79) And Julian Huxley (1953): "In all fields of inquiry, there is the danger of not seeing the wood for the trees. Nowhere is the danger greater than in evolution." (p. vi)

One thing, though. I want to make it clear that I did not set out to doggedly prove the classical theory wrong. I was ready at any one point to drop the whole project upon being confronted with experimental proof confirming the orthodox stance, which would have contradicted my informal observations. But, as I read the works of neo-Darwinists (and others) in both journals and books (and after a while it became evident that they all said the same thing), not only was this experimental data lacking, but I was surprised at first, then shocked, at the overwhelming, almost exclusive, philosophical-like *verbiage* that saturated the subject matter. In particular, a very strong antipathy began to slowly form towards neo-Darwinists with their tunnel vision and their dismissive, arrogant and condescending attitude, which often expressed itself in an anti-scientific suppression and/or misrepresentation of opposing viewpoints, compounded by their insults and even outright lies. It has gotten so bad with these neo-Darwinists that one has the urge to continue a scientific discussion with them by breaking a chair over their heads.

* * * * * * *
* * * *

For many years now---decades, really--- I had been having some vague misgivings, in bits and pieces, about the theory of evolution. Sporadically, I kept coming across facts that seemed to flatly contradict the theory.

The first was the continual recurrence of cases of humans with lethal genes (e.g., cystic fibrosis, diabetes, Tay Sachs, etc.). By all logic, these should not exist---much less keep cropping up. A Biology teacher in my university explained away the apparent contradiction by stating that an organism only had to live long enough to pass on its genes. Nevertheless, that explanation rang false. A lethal gene is maladaptive, no matter how you look at it, or what spin you put on it and the classical theory clearly and unequivocally states that any advantage that an organism has will make that organism and its descendants proliferate and supplant the original population and those that have maladaptive genes will become extinct. The logical conclusion is self-evident.

In the field of psychology, a relatively new subspecialty has cropped up, evolutionary psychology. Its adherents oftentimes give the impression of focusing on a specific behavior and then, working backwards, attempt to find an adaptationist reason for the behavior. They are always successful. It has taken such ludicrous lengths as to claim that depression and schizophrenia have been selected for through evolution because of their adaptive benefits (Andrews & Thompson (2009); Adriaens, (2007); Blease, (2015)).

This poses a conundrum for scientists who study evolution. Why would genes for such diseases exist when they decrease a person's fitness? One theory that attempts to explain this paradox is called "balancing selection"---the heritable genetic mutations that code for some diseases tend to also be beneficial in some other, unspecified, way. Richard Lewontin and John Hubby came up with the idea in 1966, positing that deleterious genes will circulate within a population to help maintain genetic diversity. Too little diversity and some individuals will suffer from deleterious diseases; just enough and some individuals will benefit from the "heterozygote," or hybrid advantage. To my knowledge, this theory has not been empirically verified and is counterintuitive.

A somewhat related example is that of animals with a specialized diet. For example, the snail kite in Florida is an endangered species. It only eats the scarce apple snails. It will not eat anything else in spite of conservationists' determined efforts into tricking it into eating other food. And if that was not enough, snail kites have the highest percentage of nest failures in birdland (66%). Its survivability hangs by the proverbial thread. The snail kite is just plain maladapted.

Stephen Gould (1980), in a bit of linguistic legerdemain, used the imperfect panda's thumb as an indication that such imperfections demonstrate evolution as an ongoing process, but such an argument cannot be used here. It's plainly absurd.

The second discordant fact was coming across in books and films examples of living fossils like the coelacanth and the horseshoe crab and others. Again, by all logic, I *felt* more than consciously *thought*, that because evolution is such an unrelenting force, their existence should have ended millions of years ago, yet . . . here they all are. But at the time I did not form the feeling into words.

Third, there were haphazard observations that nagged at me. For example, in one of my saltwater aquariums in Florida I had what is called a "chocolate chip starfish," which is a small, brown, starfish with small soft protrusions that are black, hence the name. I spent long periods of time trying to understand just what was so adaptive about its coloration that had made it evolve into its present appearance (in other words, evolved into a distinct species, different from the others), particularly in regard to its conspecifics. I was never able formulate a single reason.

Finally, the whole problem became crystallized in March 1999, shortly after having buried my mother in Orlando, Florida, as I gazed at a fossil inside a cabinet under a yellow light in a store that specialized in selling fossils. I was gazing at a perfect specimen of a marine vertebrate---a ray---with a tiny shrimp also fossilized *millions* of years ago in the same slate. That fossil instantly crystallized with a shock some thoughts---actually feelings---that had been occasionally fluttering in my head and one thing became immediately obvious and I was finally able to verbalize it: *evolution had stopped!* Otherwise, evolution would have continued developing the ray until, now, we would not have rays and would, in fact, not have recognized the organism other than being alien. The realization was like a thunderbolt from the sky.

As much as I coveted it, this was one of those periodic episodes of impoverishment in my life, and it forever passed from my grasp when someone else bought it many months later. Yet, I could not begin work on this book until much later since another misfortune overtook me that July which lasted for many years. Since then, however, my interest in fossils has strayed from the heretofore popular dinosaurs to those less appreciated, mundane fossils of the many animals that have persisted to this day after millions and millions of years of unchanged existence: crabs, sea urchins, insects, fish, rays, starfish. Compare these fossils with those from the Cambrian, some of which we have no idea which is the "head" or the "tail." (Gould, 2007)

One thing I want to make clear right from the very beginning: I do not suggest Creationism as a substitute, nor do I take it seriously (as far as I am concerned, the non-issue of Creationism vs. evolution is a stagnant bore). I mention it right from the very beginning because I have read the writings of neo-Darwinists as they immediately and arbitrarily dismiss scientific arguments critical of the theory of evolution simply because of other criticisms made by some Creationists elsewhere. Likewise, they arrogantly lump all criticism of the classical theory of evolution as being Creationism and refuse point blank to consider it. I find this despicable. And anti-scientific. I call them anti-scientific, not because of their theoretical stance, but because of their tactics. For example, Erns Mayr's arrogant response to Leon Croizant' criticism of the Darwin-Wallace theory is typical: "Time is too short to argue with such authors" (Coalacino 1977).[4]

Rational, logical, intelligent arguments and criticisms (of any theory!) should be dealt with, regardless of who makes them and not dismissed with a wave of the hand. I can understand that, emotionally, it would seem that if the Darwinian-Wallace theory is brought down, the only alternative would be Creationism, but such thinking is fallacious. One does not follow from the other, though some would like it to be so. It is *non sequitur*. The Creationists' buffoonish antics have tended to overshadow legitimate criticisms of the classical theory. Worse, it has led to guilt by association on the part of the equally close-minded neo-Darwinists.

The problem with this debate on evolution is that its participants believe themselves stuck in an either/or scenario, what I call a false dichotomy: either you adhere to the classical theory of evolution, or you are a Creationist. Neither considers the fact that there may be other rational, viable alternatives. Additionally, both sides have a tendency to believe that criticism of the *mechanism* of Natural Selection, through which evolution is reputed to take effect, equates with repudiation of the overall existence of the process of evolution; an understandable assumption, to be sure, but totally erroneous. This is a false dichotomy which has held up progress. Evolution is a given; it is the *how* that is in question in this book. Simply put, should the classical theory be conclusively proven to be erroneous, we simply have to get back to work to explain how evolution has taken place. It is really that simple.[5]

I can understand why biologists (and paleontologists for that matter) would be stubbornly resistant to questioning the classical theory. It is the only theory that they have---at least the only major theory (or meta-theory)! This is due to the centuries-old, deeply ingrained anti-Aristotelian, Baconian, axiom that fact gathering takes precedence over theorizing (Gillman, 1996). Biology is not Cosmology, wherein physicists blithely and routinely crank out one theory after another, accompanied by complex mathematical formulations that have no basis whatsoever to reality to the point that Astrophysics is in danger of becoming a fact-free science.[6,7] Consequently, the biologist holds on to the classical theory of evolution as a hermit crab holds on to its shell. Indeed, it has been said more than once that in biology nothing makes sense unless it is in the context of evolution (Conway Morris, 1998). And, really, one really has to sympathize. Biologists have precious few major theories and are blessed with a multitude of facts. This is in stark contrast to astronomers and astrophysicists who have precious few facts and are burdened with a multitude of bogus theories. Astronomers cannot even tell with absolute certainty even something as simple as whether there is water on Io, or Ganymede and, if so, how much. Yet, that does not stop them from spinning elaborate, esoteric theories of any manner, including how the universe began. Astronomers and physicists adopt and discard theories as one might underwear. At one time, or another, for example, astrophysicists confidently predicted with very elaborate, mathematical and logical arguments that: (a) Jupiter would become incinerated when it would be impacted by Comet Shoemaker-Levy 9, our solar system thereby becoming a binary (b) any spacecraft, or human, landing on the moon would be swallowed whole by an ocean of lunar dust, this dust having come into being through millions of years of being bombarded by micrometeorites (Ley, 1965) (c) a parallel universe exists (Valenkin, 2006) and (d) wormholes exist in space which will cut down the time to travel from planet to planet. The last two have been pounced upon by science-fiction writers and filmmakers in an effort to revive a dead, sterile field.[7] The latest physicists' fantasy is string theory, which Woit (2006) has rightfully criticized as being devoid of any experimental evidence and as being an untestable and, therefore, worthless theory. At any rate, one can certainly understand how protective, indeed, overprotective in some instances, biologists and paleontologists can be in respect to evolutionary theory.

Biologists, on the other hand, should look to their sister discipline for comfort. In Physics, Sir Isaac Newton's principles held true, intact, for several centuries. That is, until certain observations began to accumulate that Newton's laws could not explain, at which point the theory of relativity stepped in and, in effect, was appended to Newton's. And, then, of course, almost simultaneously, came quantum mechanics. Yet, for all practical purposes, Newton's laws are still valid. They have simply been added on to. These developments parallel the theory of evolution. Incidentally, anti-neo-Darwinists have unjustly, and absurdly, heaped scorn upon poor Charles Darwin for the anti-scientific antics and dogmatism of the neo-Darwinists. Darwin was a flexible scientist of the first caliber---nothing at all like the neo-Darwinists. It would be equally unjust, and absurd, to heap scorn upon Newton for the shortcomings of his celestial mechanics.

This book, thusly, is simply a footnote on the question of evolution. It is appropriate that I close this introduction with a quote from Barbara McClintock: "When you know you're right, you don't care what others think. You know sooner or later it will come out in the wash."

August 21, 2004

- - - - - - - -
- - - -

And then . . . nothing happened.

In a perfect world, all editors would be periodically taken out and shot and new ones would step in to replace them every five years or so.

But, unfortunately, we do not live in a perfect world.

I have published numerous papers in technical journals, a few book reviews and a handful of newspaper and magazine articles. Publication in book forms has, however, eluded me. Part of the problem, of course, is that the book publishing system is set up with a series of gatekeepers (agents, committees, etc.) whose job consists entirely of keeping out novel ideas. Nonetheless, I contacted an editor in a well-known publishing company who had previously edited a book on evolution; since evolution was a very hot topic at the time because of the brouhaha over "intelligent design," I explained that it was a perfect time for such a book. His response was that since I was not famous, he was not interested: perhaps I could make myself famous by presenting my views in scientific meetings?

All . . . right. . . .

In the next few months, I thought of trying a novel approach and submitted an abstract to the Lunar Planetary Institute, since it is primarily focused on craters and meteorites (the relevance will be seen below). It was rejected as being an inappropriate subject matter. To my chagrin, however, my (traditional) one-paragraph abstract was, nevertheless, published in their annual journal; I say chagrin because the other "abstracts" in that journal were half a page long; had I known that fact in advance I would have expounded my theory at length since abstracts in other fields are usually of one or at the most two paragraphs (don't forget in science, priority of publication is everything). I then in 2005 submitted a proposal for the forthcoming 2006 annual meeting of the AAAS in St. Louis, entitled, bluntly enough, The Theory of Evolution is Flawed. The accompanying abstract read as follows: "The Darwin-Wallace theory of evolution is examined, and the mechanism proffered by the theory to account for evolution is shown to be flawed. Among the data that indicate that Natural Selection is inadequate to account for speciation are: the immutability of numerous living fossils, the preponderance of physical traits not conducive to survival, convergent evolution, anthropomorphism of Nature by neo-Darwinists, and the limitations to selective breeding in plants and animals. In accordance with the consensus that if one piece of data contradicts a theory said theory is nullified, then it is asserted that the classical theory of evolution should be replaced by a viable one." I ultimately received a rejection months later. Since this was the first time that a paper that I had submitted to a scientific conference had ever been rejected, I was curious as to why (yet at the same time I had half expected it) so I wrote back for a reason for the rejection. I received a one-sentence reply, "This is (sic) well known that these data do not in fact contradict the theory of evolution." Frankly, the arrogance of the response and the falsity of such a claim angered me. Lastly, I wrote a note to *Nature* magazine criticizing the neo-Darwinists' overall tactics in their defense of the classical theory. It, too, was rejected.

Frustrated, I contacted the previous editor again, relating all that had happened. By then, he had lost any interest in the subject.

September 2007

The revision that you are now holding in your hands is more concise, at least I hope so. But I should also warn that the primary reason for the prolific quotes in this book has been not to use those quotes as empirical evidence, but as psychological crowbars with which to pry open the solidly shut minds. From professional and personal experience, I can categorically state that two or three persons may state the same message and it goes unnoticed, but a fourth person, by a simple rewording of the very same message results in an "Aha!" reaction that does away with the prevalent mental set. And I freely admit that the style of writing in this book has been the farthest imaginable from the usual pedantic scientific style; in fact, I am positive that for many scientists it will be tantamount to fingernails being raked across the blackboard. Hopefully, at least that will serve to wake up and shake some scientists out of their lethargy and their tunnel vision.

I also have a confession to make. I formulated this theory quite unscientifically: the basic foundation for the theory, and the objections to the classical theory, came to me upon reflection. Afterwards, upon searching the literature for support for my theory, I was naively surprised to see that others had brought up the same objections to the classical theory. I was also pleasantly surprised to read the genetic and paleontological research that essentially verified my position.

December 2017

It is with deep embarrassment that I must confess to being ignorant until very recently of Georges Cuvier's work (Rudwick, 1997). In my defense, I must point out that of all the writers of evolution that I have consulted (see references) no one else did. Therefore, I had no idea of the existence of his research. Reading his work was like hearing a voice from the past.

December 2023

FOOTNOTES TO INTRODUCTION

In my own work I usually found that, if the theory was really beautiful, its destruction was not a defeat but a victory. It led to a still more fruitful theory without doing any harm to the tangible facts uncovered by the obsolete one.
---Hans Selye, *From Dream to Discovery*

[1] The fledgling paleontologist, Edward Cope, famously inserted the skull of Elasmosaurus at the tail end, a boner that was pointed out by Othniel Marsh, thereby sparking their lifelong feud.

[2] "The proposal that an impact had killed the dinosaurs offended many, paleontologists in particular, since it came from a physicist, and an explosion of controversy followed in the 1980s." (Benton, 2003, p.97)

[3] And, for the record, no, I do not consider myself on the level of these individuals.

[4] In his diatribe against Creationists, Eldredge (2001) mentions Behe's Irreducible Complexity Principle, then dismisses it out of hand by saying that it is nothing new, we have heard it all before---then goes away without refuting it. What makes Eldredge even more contemptible and indefensible in that book is that he very deliberately, very consciously (and even states so in the book), introduces petty politics of conservative vs. liberal into the debate of evolution; for example, he weaves President Bill Clinton's impeachment into the overall scientific debate on evolution which, frankly, has absolutely no relevance whatsoever and sets a dangerous

precedent. Additionally, he was up in arms over former President Ronald Reagan having referred to the theory of evolution as a theory. Which it is. The theory of evolution *is* a theory. But evolution itself is a fact.

[5]Actually, the more I read up on the subject, the more I was surprised to be aware of scientists and other professionals' muted skepticism towards the theory, voicing some of the very same arguments that I was making, for example, this from an unnamed Dutch doctor in Africa (she was commenting on the usual argument as to why giraffes evolved to have such long necks, which was to be able to eat the top leaves of trees): "Evolution is a fact. I don't question that. Only the mechanism behind it. That eternal stretching of the giraffe's neck bores me. I want experimental evidence that natural selection is the most important evolutionary force." (Goldschmidt, 1997; p.77) Incidentally, there is enough vegetation for a camel to eat at all levels in between, including at ground level.

[6]A branch of Psychology was in danger of doing just that. When Cattell (1978) and his colleagues began to use factor analysis in order to decisively, experimentally, establish the components of personality, rather than resorting to armchair theorizing as was traditionally done, they used factor analysis (at a time when computers were bulky and time consuming). At one point they became so worried that the tedious, convoluted mathematics that they were performing had no basis in reality that they came up with the concept of plasmodes. A plasmode was an object in the real world whose dimensions could be quantified and, therefore, subjected to factor analysis. They did just that and, lo

and behold, the results corresponded to the plasmode's reality.

Flash forward thirty years later. The Cattell team (Cattell, Eber & Tatsuoka, 1970) was famous for having *empirically* researched personality traits, out of which had resulted in the widely used personality test, the 16 PF. Prior to them, personality theories were armchair endeavors, with each theorist emphasizing his own particular obsessions (e.g., Adler on feelings of inferiority and superiority, Freud on his own sexual perversions, etc.); the Cattell team took great pride in its rigorous research. Yet, when the 16 PF was finally formulated, a Motivational Distortion scale---to measure "faking good," as so many other personality tests have incorporated along with a "fake bad"---was added as an afterthought, with minimal research and with the assertion that such a scale was really unnecessary since both ends of each personality scale was equally attractive (each personality scale is bipolar, e.g., introvert/extrovert); such a statement was without empirical foundation. The Motivational Distortion scale was ultimately included in only two of the different forms of the test. Like I said, flash forward about thirty years and you have Cattell (1992) asserting that the Motivational Distortion scale is "theoretically bankrupt" in spite of the empirical fact that people who take the test with the motivation of presenting themselves in a favorable light *invariably* alter the outcome of the resulting personality profile (e.g., Jeske & Whitten, 1975; Simón, 2007).

[7]Alfven (1971), the Nobel prizewinner in his acceptance speech, made a similar observation in regard

to plasma physics:

> I think that it is evident now that in certain respects the first approach to the physics of cosmical plasmas has been a failure. It turns out that in several important cases this approach has not given even a first approximation to truth but led into dead-end streets from which we now have to turn back. The reason for this is that several of the basic concepts on which the theories are founded are not applicable to the condition prevailing in the cosmos. They are "generally accepted" by most theoreticians, they are developed with the most sophisticated mathematical methods; and it is only the plasma itself which does not "understand" how beautiful the theories are and absolutely refuse to obey them. (p. 172)

And a relevant chapter on the Chandler Wobble (Mulholland, 19, p. 238):

> It was left for an amateur astronomer and mathematician---in private life a prosperous New England merchant---to crack the puzzle. Seth C. Chandler probably didn't know enough theory to make the usual mistake. He announced in 1891 that Euler's variation was real, but the period was fourteen months instead of ten. This created a furor. As happens even in our day, the theorists were unwilling to believe an observation unless they knew a good theoretical reason for it. They were quick to claim that Chandler's analysis must be faulty or that his observations were bad. They could not imagine that Euler's theory was so wrong.

And:

> Many theoreticians sought to explain how periodic patterns [of physical development] could be organized across large structures. While the maths and models are beautiful, none of this theory has been borne out by the discoveries of

the last twenty years. (Behe, 2007, p.189)

[8] And closer to home, during the Manhattan Project, there was at least one physicist who theorized, with great logic and many formulas, that the chain reaction would continue unabated until the entire world would have disappeared. The other physicists, quite irresponsibly, went on with the famous experiment.

CHAPTER 1
THE FIRST MECHANISM: SPIRITS

If this stuff [scientific creationism] is science, why do we need a law to teach it?
---Judge William Overton

Creationists make it sound as though a "theory" is something you dreamt up after being drunk.
---Isaac Asimov

Those who cavalierly reject the theory of Evolution, as not adequately supported by fact, seem quite to forget that their own theory is supported by no facts at all. Like the majority of men who are born to a given belief, they demand the most rigorous proof of any adverse belief, but assume that their own needs none.
---Herbert Spencer

Although we do take the existence of myths and overall superstition for granted, if one stops to think about it, it is kind of odd that humans have embraced such irrational thinking. In the case of superstitions like Friday the 13th bringing bad luck, there is a psychobiological foundation. Skinner (1971, 1974) long ago demonstrated experimentally that if animals were given a pellet of food at random intervals, the animals would tend to repeat whatever behavior that had been present at the appearance of the food, even though the specific behavior was not being deliberately reinforced by the experimenter. It is not too hard to see how a belief would arise that Friday the 13th, or breaking a mirror, brings bad luck. But when you deal with elaborate myths about the world resting on the back of a giant turtle, it does seem odd---again, from a modern, rational perspective. It is particularly curious that all cultures attribute humanity's genesis to gods and their actions.

I live in a rational, industrialized society and even though we have the New Age and Wicca followers, it is still for the most part a rational society (and, in a way, that is a problem because when we come politically into contact with cultures which have a "primitive" mentality we become befuddled). My wife, who is from Indonesia, relates that the idea of ghosts, or spirits, is still very much present in her countrymen in the 21st century. Recently, she told her parents, whom she was visiting in Bandung, that she would be leaving to go to Bali; to Indonesians, visiting exotic Bali is like Americans and Japanese visiting Hawaii; her father objected strenuously as to the date she had chosen as being unlucky (but when she told him that she was taking them along, he instantly accepted and rushed out to tell all his neighbors that he was going to Bali).

Perhaps an amusing example of our modern excess rationality is when we attempt to explain the grotesque practice of circumcision, which is still widely practiced among some religious groups. The rationale that has been nowadays imposed on this grotesque custom of mutilation is that it was originally generated for hygienic purposes. Yet, if hygiene was the guiding force behind circumcision, would it not have made more sense to mandate: "Take a bath every day and wash your hands before eating"? Would this not make more sense than the mandate: "Chop off a piece of your penis"?

Millennia ago, some self-assured psychotic somewhere along the Mediterranean probably came up with the bright idea of human sacrifice; after the drought ended two or three days later, the idea stuck. We see that human sacrifices took place in ancient Greece, Carthage, the Middle East and Mesoamerica. I am certain that if human sacrifice was still being practiced today, some sophist somewhere would rationalize the practice in some way.

At any rate, this rambling preamble was simply to introduce, in a very roundabout manner, the concept that the first mechanism to explain the origin, the creation, of species was a magical one, based on the spirit world. Let us quote now from the *Popol Vuh* (actually from a translation of a translation, not that it really matters (Goetz & Morley, 1950)):

> Thus was the earth created, when it was formed by the Heart of Heaven, the Heart of Earth, as they are called who first made it fruitful, when the sky was in suspense, and the earth was submerged in water.
> So it was that they made perfect the work, when they did it after thinking and meditating upon it.
> Then they made the small wild animals, the guardians of the woods, the spirits of the mountains, the deer, the birds, pumas, jaguars, serpents, snakes, vipers, guardians of the thickets.
> And the Forefathers asked: "Shall there only be silence and calm under the trees, under the vines? It is well that hereafter there be someone to guard them."

So they said when they meditated and talked. Promptly the deer and the birds were created. Immediately they gave homes to the deer and the birds. "You, deer, shall sleep in the fields by the river bank and in the ravines. Here you shall be amongst the thicket, amongst the pasture; in the woods you shall multiply, you shall walk on four feet and they will support you. Thus be it done!" So it was they spoke.

Then they also assigned homes to the birds big and small. "You shall live in the trees and in the vines. There you shall make your nests; there you shall multiply; there you shall increase in the branches of the trees and in the vines." Thus the deer and the birds were told; they did their duty at once, and all sought their homes and their nests.

And the creation of all the four-footed animals and the birds being finished, they were told by the Creator and the Maker and the Forefathers: "Speak, cry, warble, call, speak each one according to your variety, each, according to your kind." So was it said to the deer, the birds, pumas, jaguars, and serpents.

"Speak, then, our names, praise us, your mother, your father. Invoke then, Huracan, Chipi-Calcuha, Raxa-Calcuha, the Heart of Heaven, the Heart of Earth, the Creator, the Maker, the Forefathers; speak, invoke us, adore us," they were told.

But they could not make them speak like men; they only hissed and screamed and cackled; they were unable to make words, and each screamed in a different way. (pp. 84-85)

And that is, according to religion, how all the species of animals came into being.

CHAPTER 2
THE SECOND MECHANISM: SPONTANEOUS GENERATION

"I give them experiments and they respond with speeches!"
---Louis Pasteur

Instead of going back to the facts and seeing for ourselves, we blindly follow tradition.
---Jean Henry Fabre

But science is also self-correcting: The most fundamental axioms and conclusions may be challenged; the prevailing hypotheses must survive confrontation with observation; appeals to authority are impermissible; the steps in a reasoned argument must be set out for all to see; experiments must be reproducible.
---Carl Sagan

It is always refreshing to read a contemporary account about a scientific problem investigated in the 1700s and 1800s---or even earlier---to shake us from our present world view. Ideas which we now take for granted, beyond question, even simple ones, were at one time conundrums over which heated arguments took place and all sides considered the other to be blind, stupid, or stubborn. Conversely, by the same token, it is surprising to realize that what one took to be recent knowledge or discoveries was, in actuality, known centuries before.

The concept of spontaneous generation was in existence since before Aristotle's time and lasted well into the late 1800s. It was believed that many small organisms appear, as if by magic, from the refuse of both humans and nature, without the mechanism of breeding from conspecifics. Essentially, spontaneous reaction was thought of as a sort of chemical reaction. This was not only the case for fungi, plants and infusoria, but was also the case with vertebrates like mice, fish and frogs. If a frog was seen to emerge from the slime, spontaneous generation was thought to have been at work. Mice would spring forth from grain wrapped inside a dirty shirt, insects from within trees. Rabies was thought to occur in dogs spontaneously.[1] As for parasites, they were spontaneously generated within the body (Zimmer, 2004). Illnesses were due to "spontaneous virulence" (Pasteur, 1878/1996).

No attempt was ever made to differentiate as to which animals arose from which medium. However, organic matter was a prerequisite.

The idea began to be first attacked in 1675 by an Italian scientist who wrapped meat inside gauze. Maggots were not found in the meat, but outside, on the gauze, where the flies had laid eggs. In 1745, in London, Needham, a Catholic priest, carried out experiments which seemed to confirm the phenomenon of spontaneous generation, to wit, warming nutrients in a sealed container which later showed infusoria to be present; he later came to Paris and worked with the highly influential Buffon who also became an advocate of the concept.

In 1765, Spallanzani, another Catholic priest, this

time in Modena, also using experiments, published a book attacking the concept of spontaneous generation with experiments which modified Needham's original experiments. He had repeated Needham's experiments, although he increased the temperature to the boiling point and kept it so for a long duration. Needham and Buffon retorted that Spallanzani had destroyed "the vegetative force"---whatever that was.

In 1810, Appert, in France, anticipated his later canning industry with his duplication and extension of Spallanzani's experiment, with no resulting putrefaction. Gay-Lussac, a contemporary, explained away the lack of results by Appert as being due to the lack of air; after all, all living things need air to live.

In 1836, Theodor Schwann, in Berlin, boiled meat broth in a glass flask, sealed it and allowed cooled air to be introduced which had then been first greatly heated. No putrefaction of the meat broth took place and Schwann made the canny observation that the process of putrefaction was similar to that of fermentation.

He really should have followed up that observation.

This should have been the end of it. But, in spite of these isolated experiments, belief in heterogenesis stubbornly persisted and it should be pointed out that, ironically, the increased widespread use of the microscope added life (no pun intended) to the argument of spontaneous generation since microorganisms could now be clearly seen.

In 1860, France's Academy of Sciences offered a prize to---for all intents and purposes---solve the problem once and for all. Louis Pasteur went to work on it, as did his opponent, Pouchet, though the former saw it as a digression from his main work on fermentation.

At this point in time Pasteur had a number of scientific successes under his belt, but nothing compared to what was to come from 1860 onward (Sigerist, 1971). One can see that although the topics of his impressive, lifelong, investigations varied greatly (anthrax, rabies, gangrene, spontaneous generation, silkworm diseases, vinegar, fermentation), all were actually variations upon a central principle: microorganisms break down organic matter to the point of putrefaction. Once Pasteur himself had realized this principle, he said that he became "obsessed" with the idea that bacteria putrefied healthy bodies, i.e., disease.

> These new studies are based on the same principles which guided me in my researches on wine, vinegar, and the silkworm disease---principles, the applications of which are practically unlimited. The etiology of contagious diseases may, perhaps, receive from them an unexpected light. (Pasteur, 1879/1996; p. 15)

Another thing that is remarkable about Pasteur is that once he solved a problem of great importance he did not rest on his laurels to receive praise, as had/has been customary, but went on to the next problem and the next one. He was, in every sense, a modern scientist.

Very briefly, these were some of his experiments on spontaneous generation, which stretched for a number of years.

First, he trapped airborne particles in cotton asbestos and examined them under a microscope and found germs to be present.

Secondly, he did the traditional experiment of heating, then sealing, a flask containing a liquid nutrient. After six weeks, the cotton's particles were introduced into the nutrient. By the end of the second day, putrefaction had set in, indicating that a) the nutrient did not lose its "vegetative force" through use of this method b) germs existed in suspension in the air c) the same organisms grew in the nutrient as would ordinarily. His previous work on fermentation had shown the same thing: if all contact with air is cut off, fermentation does not occur.

Third, he put heated nutrients in flasks whose unsealed necks were greatly elongated and twisted (i.e., swan neck retorts), so that airborne dust could not easily navigate through the necks. No putrefaction occurred (I understand that the original flasks with their original, unspoiled, contents are still at the Pasteur Institute).

Fourth, and possibly his most dramatic experiment, often told to schoolchildren, during Pasteur's vacation time, he traveled to the Alps with a number of sealed flasks. At different altitudes, he opened the flasks with a heated instrument, then resealed the flasks and found that, as the elevation increased, there was a decreased probability of contamination. Germs may be present, but they are not ubiquitous.

Pouchet refused to believe that the air contained microorganisms. Quite logically, he objected that if germs were, indeed, airborne, then it would only be a matter of time until they formed a thick fog, to which Pasteur pointed out that they needed nutrients in order to multiply. Pouchet, incidentally, during this time, had been carrying out experiments of his own which he claimed confirmed heterogenesis. However, Pasteur examined them and pointed out the experimental errors that Pouchet had inadvertently committed. A committee by the Academy of Science awarded Pasteur the prize on default: Pouchet had withdrawn in protest to all the increasingly rigid restrictions insisted upon. For a few years thereafter, adherents of spontaneous generation still persisted, but eventually died out.

What is truly curious, however, is that, initially, Pasteur *was* looking for spontaneous generation and was searching for it, but his systematic, detailed, controlled experimentation led him to its negation (Nicolle, 1961).

Incidentally, politics intruded itself into the scientific question, with French liberals advocating spontaneous generation and conservatives against it. It is reported that the question was argued with much passion at the time. Why? Well, if spontaneous generation did take place, then God was not involved at all, and one did not have to invoke Him to account for the origin of species (including humans). God was out of the equation. This is why the number of papers written on heterogenesis saw a sharp increase during this time (Debre, 1994). Let us see what Pouchet himself had to say about this:

> There are those people who consider the whole question of spontaneous generation an impudent challenge thrown in the face of religion, a most dangerous question liable to sap the foundations of our beliefs and overthrow the laws of Earth and Heaven. Such fears are unfounded, for if this phenomenon exists, it is because God has desired to employ it for His own ends. (Nicolle, 1961, p.60)

As is often the case, the controversy made Pasteur famous throughout France and brought him to the attention of Emperor Napoleon III (one would cynically expect that the Emperor would have sided with Pasteur's opponents, but this was not the case, even though the Emperor had sent troops to safeguard the Pope in Rome). Pasteur actually met the Emperor several times and brought along with him his microscope and some other equipment and the two of them, accompanied by Napoleon's wife, enthusiastically carried out a number of experiments (Benz, 1938).[2]

When all is said and done, spontaneous generation was another, fancier, term for magic. Can life be created out of nothing? Yes. How did species come into being? They just did. The scientific proponents of that theory attempted to couch it in scientific terms and experimentation, but the basis for it was, essentially, magic.

Incidentally, in the twentieth century we saw a similar thing with psi phenomena, with its technical jargon, the faulty experimentation, the experimenter bias (e.g., Rhine, 1975). At the end of the day, "ESP" (and its subdivisions of telekinesis, clairvoyance, etc.) is just another word for "magic," except that the word "magic" can simply no longer be taken seriously. Even professional magicians like David Copperfield and David Blaine do not call themselves Magicians, they call themselves Illusionists.

There must be a strong tendency within human psychology for magic. We can watch Hollywood movies on ESP with fear, tension and awe instead of laughing at them because of the obvious absurdities that they portray. Yes, the tendency to believe in magic is strong, even if we are not consciously aware of it. Nor does it end with ESP. Even modern-day scientists are prone to rely on spontaneous generation:

> The finest example is Meteor Crater in Arizona, where Gene [Shoemaker] had found really convincing evidence for an impact origin. Conventional geologic opinion attributed these craters to mysterious explosions that occurred at random times and places for no evident reason. In retrospect this causeless mechanism for making craters is indistinguishable from magic, but at the time many geologists considered it preferable to catastrophic impacts. (Alvarez, 1997; p.76)

How did life first originate on Earth? Chemicals came together to form the first amino acids which became self-replicating.

In other words, spontaneous generation.[3]

How did the Universe come into being? There was the Big Bang and suddenly the Universe was filled with galaxies and stars.

In other words, spontaneous generation.

FOOTNOTES FOR CHAPTER 2

Fortunately, it is not so much what we do not know that handicaps our research, but what we think we know, although it is false.
---Hans Selye, *From Dream to Discovery*

[1] I suppose that one could argue that Mary Shelley's *Frankenstein* has as its basis heterogenesis, in so far as the authoress was acquainted with some of the scientific events of her time, as she herself wrote in her 1817 Preface: "The event on which this fiction is founded has been supposed, by Dr. [Erasmus] Darwin and some of the physiological writers of Germany, as not of impossible occurrence" (Shelley, 1818; Kemp, 1998).

[2] I know that I am going to be misunderstood for the following, but I have to confess that I have often felt sorry for Napoleon III. At least in the way that his countrymen have viewed him, he has been the subject of much undeserved abuse. The megalomaniac Napoleon I is much admired, inside and outside of France, for having turned parts of Europe into a charnel house and for stealing anything that was not nailed down and sent to Paris. Napoleon I's military genius is self-evident but, in the end, for all of the blood spilled, France's borders remained unchanged. He was, essentially, a directionless, compulsive, warmonger. Napoleon III, on the other hand, permanently increased France's borders with a minimum of bloodshed---certainly no more than

was typical of the times---he improved the economy, made numerous, lasting, internal improvements, such as founding scientific laboratories, broadening the medieval, unsanitary, Parisian streets (which French cynics typically dismissed as being for the purpose of better troop movements to quell uprisings). Furthermore, a sensitive man, he lost whatever taste for war that he might personally have had because of his name when he saw the butchered bodies after a battlefield.

> [3]"The third alternative [hypothesis explaining the origin of life on this planet], that living substance evolved out of nonliving, is the only hypothesis consistent with scientific continuity. The fact that spontaneous generation does not occur now is no evidence that it did not do so at some earlier stage in the development of this planet, when conditions in the cosmic test tube were extremely different. Above all, bacteria were not then present, ready to break down any complex substances as soon as formed.
> (Huxley, 1953; p. 21)

CHAPTER 3
THE THIRD MECHANISM: LAMARCK

One is forced to recognize that the totality of existing animals constitute a series of groups forming a true chain, and that there exists from one end to the other of this chain a gradual modification in the structure of the animals composing it, as also a proportionate diminution in the number of faculties of these animals from the highest to the lowest (the first germs), these being without doubt the form with which nature began, with the aid of much time and favorable circumstances, to form all the others.
 --- Jean-Baptiste Lamarck

The sick in soul insist that it is humanity that is sick, and they are the surgeons to operate on it. They want to turn the world into a sickroom. And once they get humanity strapped to the operating table, they operate on it with an ax.
 ---Eric Hoffer, *The Passionate State of Mind*

The inheritance of acquired characters, occasionally referred to as "Lamarckism," is a scientific theory of evolution, one of the earliest, if not the earliest, published in 1809. It states that an organism, in encountering an environmental difficulty for which it has to cope will, indeed, cope and its descendants---as a *direct* result of the parental efforts---will be better able to cope because they will have acquired those physical attributes that help them cope. Indeed, one finds that an organism will vary in morphology depending on what environment it finds itself in. The clichéd example that is always given is that of an herbivore, in trying to reach leaves in the branches of trees, continually stretches its neck to get at them. Through time, its descendants acquire a longer and longer neck until we have the "cameleopard," as the giraffe was known in the 1800s (it is always forgotten that a giraffe can also eat grass). In fact, human beings have proven adept at doing the very same thing in their livestock, in their pets and in their crops and horticulture. There are many varieties (or races) of cattle, horses, dogs, roses, orchids, and the like as different from the original prototype as night from day.

Furthermore, since members of a species and their descendants are constantly changing because of the differing environment, it is a mistake to say that there is such a thing as a "species;" for Lamarck, species had indefinite powers of modifying their morphology in order to adapt to changing circumstances.

> Thus, among living bodies, nature, as I have already said, offers only in an absolute way individuals which succeed each other genetically, and which descend one from the other. So the species among them are only relative, and only temporary. Nevertheless, to facilitate the study and the knowledge of so many different bodies it is useful to give the name of species to the entire collection of individuals which are alike, which reproduction perpetuates in the same condition as long as the conditions of their situation do not change enough to make their habits, their character, and their form vary. (Lamarck, quoted in Packard, 1901)

Ironically, some radical neo-Darwinists have said the very same thing, that because evolution is going on all the time, which is changing the morphology of organisms and their descendants, it is a mistake to say that there is such a thing as a "species."

Additionally, he argued that if one looks at Nature and the fossil record, one sees in species the tendency towards *progressive* advancement towards a more perfect morphology, going, that is, from the simple to the complex in morphology.

Jean-Baptiste Lamarck was a French scientist who worked in the early 1800s. He wrote on various fields such as geology, botany, taxonomy and paleontology. In 1809, he put forth his theory of evolution; his work on the evolution of species went completely ignored until noticed much later, particularly by those outside of France (aside from anticipating the theory of evolution he also anticipated Malthus). He was one of those rare prominent individuals in the history of science that we come across from time to time who, although they generate erroneous ideas, at the same time they also generate a respectable amount of verifiable, scientific work.[1] Additionally, his personality is reputed to have been . . . odd. However, the negative aura around Lamarck, according to Packard (1901) and Schwartz (1999), is actually derived from Cuvier's spiteful "eulogy" (although a translation of the eulogy lacks any vitriol (Cuvier, 1836)).

To us in the present, Lamarck's name has survived primarily by being attached to the concept of the inheritance of acquired characters to the point that the idea has also been called Lamarckism. However, he did so in such a manner that he brought disrepute to the concept of evolution, hence Darwin's exclamation (presumably made before his pangenesis amendment to his previous theory of evolution, which was an obvious copy of Lamarck). Nonetheless, for the sake of simplicity we will henceforth refer to the concept of the inheritance of acquired characters as Lamarck's. It should be noted that a very, vague idea that present-day animals were to some degree descended had been floating around Western Europe even prior to Lamarck (e.g., Buffon). After all, if one looks at Linnaeus' taxonomy, one becomes aware of the fact that there is an increasing (ascending) complexity in animals.

The Lamarckian idea, unquestionably, has a strong intuitive appeal. In fact, we find that it sometimes indirectly and unintentionally creeps into modern parlance by both laymen and scientists; a couple of examples involving insectivorous plants (beginning in each sentence insert the words, "in order"): "To capture their prey, the plants have developed three different trap styles." (Schwartz, 1974, p.9) And: "To adapt to these environments deficient in many minerals and possibly in some cases to overcome the inability of their root systems to absorb required minerals, some plants retained the evolved capacity to trap and digest small animals." (Schnell, 1976, p.2) Elsewhere, Zimmer (2001, p. 202) comments on the rise of mankind, "This abstract thought made it possible to make better tools, and survival became even easier. Tools, in other words, may have made our brains swell." Even a neo-Darwinian like Richard Dawkins (1986) slips into problem solving. Writing of flounders and soles: "But this raised the problem that one eye was always looking down into the sand and was effectively useless. In evolution this problem was solved by the lower eye 'moving' round to the upper side." (p. 92) And Ernst Mayr (2001) strikes a suspicious sounding Lamarckian tone:

> The thecodont ancestors of birds were close relatives of the ancestors of the dinosaurs and can be assumed to have had a rather similar genotype as the dinosaurs. The shift to bipedal locomotion may have induced their similar genetic endowment to respond with a similar morphological construction as the bipedal birds. (pp.67-68)

Such is the intuitive appeal of the idea of inherited acquired characteristics that we find individuals in other cultures adhering to the idea although they may have never heard of Lamarck or have been informally introduced to the idea. Admiral Yamamoto, for example, had two fingers amputated as a result of an injury in a battle; when his son was born, he asked the midwife whether his son had all five fingers in each hand (Hirokuyi, 1979).

At any rate, there are apparently several problems with this particular theory of evolution, which can be characterized as Evolution Through Willpower. The first, simply, is the lack of solid evidence. The second is the lack of a mechanism for handing down acquired traits to the descendants; the process of DNA-nucleic acids to proteins to phenotypes is a one-way street. The third is plants. Let us elaborate.

First, if Lamarck's theory was correct, then we would see that the offsprings of plants who had been subjected to topiary would exhibit growth in the form of seals with balls, elephants, or Mickey Mouse. Likewise in those populations where the grotesque mutilation of circumcision consistently takes place (Jewish and Muslim), we would see baby boys coming into this world already circumcised. Of relevance, in a famous experiment, August Weissmann, put the Lamarckian theory to the test by cutting off the tails of mice for 22 generations to determine whether the offsprings would be bobtailed. The experiment had negative results (Koestler, 1971). The same "experiment" has traditionally been carried out with Doberman dogs: puppies have normal tails which are subsequently amputated for who knows what reasons, yet Dobermans continue to be born with normal tails.

But perhaps the objection is made that for Lamarckism to take effect, the organism must exert itself, i.e., there must be a behavioral effort. Yet, we see that human infants are not speaking earlier and earlier. Also, again, we see that body builders do not necessarily have descendants who come into the world with a flawless physique. Indeed, if one considers the fact that, for millennia, men had to rely on raw, brute force to get things done---only recently just having to merely push buttons to get those same things done---if Lamarckism was correct, then every man in the world would automatically have the physique of an Arnold Schwazenegger or Dwayne Johnson and we can clearly see that that, unfortunately, is not the case.

As for the second objection, the absence of a precise mechanism within the cells for passing on acquired traits, in all fairness, Darwin and Wallace were likewise constrained by the lack of the precise process for the inheritance of those new, adaptive, traits that had been selected for. That constraint vanished with the discovery of Gregor Mendel's neglected work (it is often forgotten that in the late 1700s and early 1800s, the sciences were in their embryonic stage; the greatest minds of that period were groping in the dark). However, no biochemical or genetic process has, as of yet, been found that would buttress the inheritance of acquired traits theory of evolution---after two centuries since Lamarck made his suggestion.

The third objection, made by Darwin (Gale, 1982), is that the theory leaves out the entire Plant Kingdom. Evolution through willpower, as a concept, works up to a certain point with plants in their trying to grow higher in order to receive more sunshine (hence we have trees). But that is all. Does a plant "will" to grow red instead of yellow flowers, or large fruit, or large seeds, or have pinnate, dentate, or orbicular leaves? Does a plant suddenly "decide" to become insectivorous and begins to modify its leaves in order to form a trap with which to trap insects?

Fourthly, another shortcoming of this hypothesis is that it does not explain passive characteristics in animals. Does the robin "will" to have a red breast, or the male cardinal "will" to be completely red with the exception of the eyes, which are black? Improbable. Or even traits that will later be active? Does a plant decide and "will" that it will become carnivorous and thusly, its unremarkable, passive leaf becomes a pitcher in which insects will fall in and be digested, or a sundew changes its leaf so that sticky tentacles form and can trap, enfold

and digest gnats? Unlikely.
EPIGENESIS
Having said the above, some present-day scientists nevertheless hold the opinion that Lamarck's ideas are not entirely dead and, what is more importantly, there are impressive experiments to back them up, so that genetic determinism is not iron clad. Indeed, there is a cluster of experiments along these lines that is truly exciting, together referred to as epigenetics (Hall, 1998; Rassoulzadegan, et al., 2006; Richards, 2006; Watters, 2006; Schmidt, 2007; young, 2008; Eakin, 2014; Skinner, 2017), which is the inheritance of physical traits that do not depend on variations of DNA.

Burr, Hyman and Myers (2001) are of the opinion that HIV's origin must be due to a Lamarckian process. Steele's (1999) and Steele, Lindley and Blanden (1998)'s research leads them to conclude that details of acquired immunities to disease by parents can be passed on to their offsprings, circumventing Weismann's Barrier ("if acquired inheritance in the immune system is not a real phenomenon then the only *ad hoc* alternative would be to invoke an intelligent gene manipulator, or 'divine interventer' as playing a role in evolution" (p.22)). Additionally, Paramecium whose cilia is surgically removed and reimplanted in the opposite direction will pass off that characteristic to its offsprings (Margulis, 1998). Nutrition also plays a role (Cropley, Suter, Beckman & Martin (2206; Fraga, *et.al.,* 2006). For example, Waterland & Jirtle (2003) fed folic acid and other methylene-rich supplements to pregnant agouti mice, which had offsprings with brown fur as opposed to those pregnant mice without the supplements whose offsprings developed yellow fur and a higher probability of cancer, diabetes and obesity; these differences persisted through generations, independent of nutrition. In another study (Anway, Cupp, Uzumku, & Skinner (2005), gestating rats were given chemicals that disrupt endocrine function; this led to offsprings with fertility defects which persisted in subsequent generations that were not administered the chemicals. Another study looked at generational transfer of irritable bowel syndrome (Theodorou, 2013). Such studies have led to the suggestion that a new subspecialty should come into being, "nutritional epigenetics." The role of substance abuse in epigenetics should be the next logical step.

Equally interesting, apart from the transgeneration physical effects, it has been found that *behaviors* in animals, specifically learning, can likewise be affected, then passed down several generations (Dell & Rose, 1987; Weaver, et al., 2004; Dias & Ressler, 2014; Nätt, et al., 2017; Kwon, 2018; McCarthy, et al., 2018). This is tangentially reminiscent of McConnell's planaria experiments.

Epigenetic studies in plants have also taken place (Lira-Medeiros, et al., 2010; Jacobsen & Meyerowitz 2007; Pennisi, 2013; Cortijo, 2014). In fact, the phenomenon was known in plants long before it was proven in animals, the latter being what caused the sensation in scientific circles (Heard & Martienssen. 2014). In this area, it is interesting that Barbara McClintock recognized the effects of transposons in maize.

So far, epigenetic inheritance has been documented in 42 species: in fungi, protists, plants and animals (Jablonka & Lamb, 2010).

And the cytological mechanism for this phenomenon? Chemical groups, such as the methyl molecules, attach to DNA and either inhibit or prevent the expression of genes (the mechanisms are actually extremely complex (Pearson, 2006) to the point that the whole concept is as mindboggling as trying to get a grip on the fact of intergalactic space).

Just as interesting as the fact, and the how, of transgenerational epigenetic inheritance, is the fact, and the how, of organisms reverting back to the original state after a few generations is also interesting. Some are relatively transient whereas others are multigenerational (Jablonka & Lamb, 2010). Additionally, epigenetic changes have been detected in ancient human genomes (Orlando & Willerslev, 2014). Epigenesis may even play a role in cancer (Suvà, Riggi & Bernstein, 2013).

(What is of particular interest in regard to our discussion here is that, whereas Natural Selection works on an individual basis, epigenesist can be a group process.   )

Neilson (2017) has suggested that the stress of centuries of dealing with natural disasters have affected the Japanese people while Rohner, et al., (2013) found that extreme low salinity results in cave fish losing their eyes, as opposed to the surface waters.

Nonetheless, it is highly amusing seeing such a noted neo-Darwinist as Richard Dawkins grasping at straws in trying to incorporate these Lamarckian researches into the classical paradigm (Dawkins, 1982); his desperation brings to mind the epicycles that were attached to the Ptolemaic system. "I am heartily sick of the epigenetics bandwagon!"

BEHAVIORISM

Lamarck's thesis was, in a sense, solidly passed on into the 20th century, though in highly mutated forms.

In the United States, it made its initial presence known in the embryonic field of Psychology near the 1920s through Behaviorism (true, it was also an extension of John Locke's *tabula rasa* concept). Behaviorism arose through the efforts of one man, J. B. Watson (1924), in reaction against the rampant unscientific mysticism and anthropomorphizing that was going on in the field of Psychology, coupled with the obscene rantings of Sigmund Freud who, incidentally, was a fervent believer in Lamarckian evolution (Odajnyk, 1976). As such, it brought much needed criticism and was greatly helpful in setting scientific standards. Watson's claim, that all behavior could be explained away in terms of Pavlovian Classical Conditioning, including learning to speak, however, stretched credulity to the breaking point. Watson, incidentally, was a fervent admirer of Paul Kammerer, a contemporaneous, well-known Lamarckian (Koestler, 1971).

Radical Behaviorism got its second wind with B. F. Skinner (1974) in the 1950s when he formalized the principles of Operant Conditioning, which later on came to have very useful applications. Nonetheless, the core radical Behaviorist argument remained that all behavior was learned (from the environment), that Pavlovian Conditioning and Operant Conditioning were the basis for all behavior in both humans and animals. Neither genetics, nor instincts, nor physiology had any influence in a person, or in an animal's behavior. Vague concepts like "personality," "dreams," "introspection," "mind" and "thinking" were sneered away as being immeasurable, untested, untouchable, and therefore unverifiable; some Behaviorists particularly liked to irritate their colleagues (and their students) no end by saying that there was no such thing as thoughts, mind, personality, or dreams.

Skinner (1974), incidentally, struck a very Lamarckian tone with the following:

> The same may be said of operant reinforcers. Salt and sugar are critical requirements,and individuals who were especially likely to be reinforced by them have more effectively learned and remembered where and how to get them and have therefore been more likely to survive and transmit this susceptibility to the species. (p.47)

Around the late 1960s, A. Bandura showed that still a third type of learning, Modeling (i.e., imitation), was experimentally verifiable; his well-publicized experiments showed that children who watched violent individuals behaved violently shortly afterwards. Behaviorists now claimed that Classical Conditioning, Operant Conditioning and Modeling were the basis for all behavior in animals and humans.

Even at the time it was widely recognized that Behaviorism had for many decades placed a stranglehold in the field of psychology, though it was by no means absolute. Its outlook was narrow, its effect inhibiting, and graduate students were steered away by Behaviorist professors from disreputable topics. Nonetheless, Behaviorism did a real service to Psychology. Behaviorists were absolutely adamant that Psychology, if it was to gain any respect, any solid standing, must become strictly scientific and experimentation must underlie all assertions. They set up rigid standards, often not so much as to methodology as to interpretation of results. They were the intellectual pit bulls of the field and a professional argument with them was nothing to sneeze at; no matter what experiments one might bring to their attention (like Harry Harlow's (1953, 1970) landmark experiments), they somehow always managed to cut the ground from under one's feet. They were truly superb debaters (if the old-style Behaviorists were around today to hear the fatuous arguments of the neo-Darwinists, the Behaviorists would tear them to pieces and eat them for lunch---followed by eating the Creationists for dessert). My graduate professor at Wichita State University, G. Y. Kenyon, was just such a one, a thin man whose pointed nose and shock of white hair combed straight back always reminded me of an older Woody Woodpecker; he was one such Behaviorist and he was the type of individual that you used to meet in universities, unlike today, with whom you could strongly disagree on an intellectual level, to the point of frustration, yet retain an immense respect for their razor sharp minds; he would often recommend you to read a published article---it wasn't much of a paper, but, boy, it had some great references!

Completely contrary to what one would have expected from the 20th century, the Behaviorists were apolitical. In spite of their rigid claim that all behavior was the result of the environment's effect on the individual through learning, totalitarian leftists and Behaviorists did not seek each other out, all the more surprising when one considers that B. F. Skinner was on the faculty at Harvard. Indeed, upon Skinner (1971) publishing his *Beyond Freedom and Dignity*, he had the distinguished claim to his credit that of being attacked by both the political left and the right (e.g., Machan, 1974).

Ultimately, the Behaviorists' stranglehold on Psychology ended towards the beginning of the Seventies, when their colleagues, instead of trying to convince them, simply bypassed them and went on to study "forbidden" topics (the birth of the specialty of Cognitive Psychology was one result). Simultaneously, European ethologists appeared on the scene carting mountains of data which became embarrassingly difficult to explain away simply in terms of learning (one can easily detect in the last writings of Skinner (1974) a considerable mellowing out, due, no doubt, to the times---the environment). Yet, I say again, for all their many shortcomings, they helped to keep the focus on a solid scientific basis for Psychology at a crucial time, when there was very little material to work with (as I was proofreading this chapter, an article in *Nature* came my way which stated that now that Behaviorism is relegated to the corner, anthropomorphism is acknowledged to be once again rearing its ugly head (Wynn, 2004)).

LYSENKOISM

The other, more direct, 20th century mutation of Lamarck came in the form of Marxism through the machinations of T. D. Lysenko, one of those vicious mediocrities that infested every corner of the globe throughout the 20th century and caused so much havoc (and continue to do so to this day).

First, let me quote Richard Feynman (2202), because his observations will be relevant:

> I would like to remark, in passing, since the word "atheism" is so closely connected with "communism," that the communist views are the antithesis of the scientific, in the sense that in communism the answers are given to all the questions---political questions as well as moral ones---without discussion and without doubt. The scientific viewpoint is the exact opposite of this; that is, all questions must be doubted and discussed, we must argue everything out---observe things, check them, and so change them. The democratic government is much closer to this idea, because there is discussion and a chance of modification. One doesn't launch the ship in a definite direction. It is true that if you have a tyranny of ideas, so that you know exactly what has to be true, you act very decisively, and it looks good---for a while. But soon the ship is heading in the wrong direction, and no one can modify the direction anymore. So the uncertainties of life in a democracy are, I think, much more consistent with science. (p.511) [2]

One of the tenets of Marxism was that there are no inherent physical and mental differences between social classes when it adopted the 18th century philosophy that had been floating around that all men were equal; the idea that this also applied to women came much later in the 20th century (and, so far, to date, has not reached at all in the many backward Muslim countries). As with Behaviorism, the environment dictates the human being, heredity being irrelevant.

As so often happens in all totalitarian systems, the political principles of the Union of Soviet Socialist Republics were extended to apply to all branches of society, so that it became commonplace to have a Marxist astronomy, Marxist art---called "Soviet Realism" (Bown, 1991)---Marxist agriculture, Marxist psychology (Winn, 1962), Marxist biology and Marxist history (in the West, we still have several diehard totalitarians in universities either teaching Marxist history, or, the field of Sociology itself considers Marx as a legitimate source of inspiration).[3] What made these inroads more damaging was that Marx, Hegel, Stalin and Lenin, considering themselves to be universal geniuses, rendered opinions in writing about various professional fields and/or specific subjects. Once firmly in power, the Communists implemented these opinions at the point of a gun.

As an aside, let me say that Arthur Koestler (1965), an ex-Communist himself, got it right and the rest of us all---with the notable exception of Conrad Zinkler, who has been grossly overlooked---completely missed the boat when he was the first to point out that Marxism, in many ways, is a religion. I do not wish to go further off on another tangent again, but let me simply say that, in hindsight, I agree with his assessment: God is replaced by History. Fanaticism is present in both. An Inquisition is present. Censorship is present. In both instances, present suffering must be accepted for a future reward (Paradise, or, a Socialist Future for the New

Soviet Man). But, most relevant to this section, is that there exists in both cases Sacred Writings which must not ever be questioned and must be forcefully applied and consulted---actually, quoted---to resolve any dispute. Heretics must recant and/or be executed.

The fascinating, if not sickening, case of Lysenko has been detailed by several individuals, both in the West (Huxley, 1949; Conquest, 1968; Zirkle, 1959; Graham, 2016) and in the Soviet Union itself (Heller & Nekrich, 1986; Medvedev, 1970; Medvedev, 1971; Medvedev & Medvedev, 1971; Sudoplatov, 1994; Pringle, 2008), although "the Lysenko affair," as it is euphemistically referred to in the West, is not as well-known as it should be---and I cannot strongly enough urge the reader if he/she reads one other book this year to read in full at least one of the above references to get all the grim details.[4]

Here, then, in a nutshell.

The science of genetics was very advanced in Russia, definitely on a par with the most advanced institutes in the West, partly due to the brilliant work of Vavilov. Unfortunately, every aspect of society was constantly scrutinized for the slightest---usually imagined---hint of enemies. Mendel, and particularly Malthus, had been demonized by Marxists and their theses dismissed out of hand; no doubt, the aversion was exacerbated by both of them being clergymen. Lysenko, together with his vicious crony, I. I. Prezent, clawed their way up through a judicious use of slogans and deadly accusations of political heresy. In the Soviet Union, an accusation, even an anonymous one, was an automatic elimination. In 1935, he came to the personal attention of Stalin, who fostered his career (Stalin saw in Lysenko a kindred soul, both from a simple, backward, background looked down on by the arrogant well-educated. Both were resentful). He became president of the Lenin All-Union Academy of Agricultural Sciences. In 1948, he set a trap in announcing a public debate on the topic of heredity in species within a session of the L.A.A.A.S. In the session, he put forth his theory, then the geneticists were allowed their say. They tore into him, and they cited experiment after experiment as the basis for their views. The Lysenkoists offered not experiments, but their usual verbiage and quotes from Marx, Engels, Stalin and Lenin. After the geneticists had had their say and had let all known their positions on the subject, Lysenko let the cat out of the bag and announced that his theories had been officially approved by the Central Committee (those poor scientists must have lost all color when they heard the announcement). In the months to come, scientific disagreement with Lysenko's theories were settled in tried-and-true Marxist fashion: all opponents were executed, or sent to concentration camps in the Arctic Circle, their laboratories smashed, and their papers burned. For all practical purposes, genetics as a science ceased to exist in Communist countries. An entire science was decimated. Lysenko's predominance[5] lasted all the way into the Sixties (incidentally, when H. J. Muller heard of what happened, he resigned his membership in the USSR Academy of Science in protest). Fraudulent experiments became commonplace.

And what were Comrade Lysenko's theories? Simply that neither genes nor chromosomes contained hereditary substances of any sort whatsoever, that heredity was not internal, that mutations did not occur, that genetic discoveries made in Great Britain, the United States, France, Spain, Italy, or Germany were to be discredited simply because of where they were carried out, that a plant's morphology---or, if you will, its species assignation---was entirely a product of the environment, also that Lamarck was acknowledged as a distant source of authority, though, of course, not on the level of "titanic geniuses" like Lenin, Marx, Engels, or Stalin. Additionally, although Lysenkoists admitted to interspecies competition (read: classes), they rejected intraspecies conflicts. Lysenko claimed that vernalization could become hereditary (Graham, 2016). Some Lysenkoists claimed to have transformed one plant species into another; presumably, they could have transformed pineapple plants into palm trees and pine trees, or strawberry plants into poison ivy. A genetics professor of mine once visited the Soviet Union, I believe in the early 1960s, and was given a tour of these fields by Lysenko himself; very puzzled, the visitor asked him, "But, where are your controls?" "Ah," replied Lysenko, lifting a finger in all seriousness, "That's one of our techniques: we don't use controls!" In short, the science of genetics vanished.[6] Few Westerners are aware, however, that, notwithstanding their own theories, it was an iron clad rule in the Soviet Union that the grandchildren/nephews/nieces and the great-grandchildren/ nephews/nieces, etc. of the aristocrats and middle-class members of 1918 were to be considered unreliable and were systematically discriminated against. Counterrevolutionary attitudes, apparently *were* hereditary.

Of some relevance to today, one of the arguments (never experiments!) used by the Lysenkoists against the geneticists was the following: the concept of fixed traits being inherited, not from the environment, but from an organism's lineage smacks of racism and Fascism, particularly the Nazi ideology (as the reader must undoubtedly be aware, that kind of argument is still very much used today in the West regarding various subjects). To illustrate the mentality involved: an architect was shot in the Soviet Union when it was discovered that, viewed from the air, the building that he had designed and constructed, by coincidence, looked like half of a swastika (Conquest, 1968). In many French, Dutch, British, German and American universities, you easily find the same mentality today.

The phrase "survival of the fittest" furthermore irritated Communists' sensibilities. They had tender sensibilities. They tended to be volubly lachrymose at the plight of The People while simultaneously turning their population into slaves and---literally---starving them to death.

Obviously, it is totally unfair to saddle Lamarck with the excesses of the 20th century totalitarians, just as it is unfair as to saddle Darwin with Nazism (which has been done). These examples are used to point to the underlying philosophy that the environment completely shapes the organism, regardless of heredity. It behooves some scholar somewhere to trace the intellectual paper trail between Lamarck and the Lysenkoists---and the numerous and various other Stalinists in the West.[7]

The argument of nature vs. nurture, in both

phenotype *and* behavioral traits, for example, also became politicized in the West by the Stalinists. Although they have backed off from claiming that the phenotypes are due to the environment in the past few decades, they simultaneously became even more militant, if not downright venomous, in the behavioral arena where they felt that human behavior was infinitely malleable.[8] Let us look at three examples (out of many).

First is the radical feminist movement of the 1970s through the present. Although the feminist movement---which called itself the Women's Movement, even though many women strongly disagree with many of their principles (Sommers, 1994; Paglia, 1994; Fox-Genovese, 1996)---started out to correct undeniably unjust situations (one of the most obvious being demanding equal pay for equal work), there was a Stalinist element within it which soon became dominant. This is not too surprising when one considers that the movement arose directly from the Marxist terrorist groups of the 1970s such as the S. D. S. and the Weathermen (again, one of those things that we are not supposed to mention in universities). At any rate, the history of the movement---which has been sanitized for modern consumption---shows that some feminists got their academic degrees for claiming that, for example, Beethoven's Fifth Symphony was a paean for rape and that Newton's *Principia Matematica*, full of mathematical equations, was in reality a rape manual, that $E=MC^2$ is a sexist equation, all of which sound surprisingly similar (changing just a few words) during the height of Soviet Stalinism. A less esoteric doctrine and more down to the earth was the dogma that menstrual discomfort in women was a conspiratorial brainwashing by men in order to keep women in their place. At any rate, the relevant point is that one of the aggressive dogmas put forth during that time was that there was no difference---*none whatsoever* and I am neither exaggerating nor inventing this, as anyone who lived during that time will testify, they really meant it---between men and women. None! Anyone who laughed at that assertion was pounced upon verbally and quite simply, and typically, overwhelmed with shrieks and abuse. Furthermore, an unappetizing physical-and-mental ideal was put forth, called "androgyny" which, for some reason, never caught on.[9] Although I am personally in strong agreement that there should be no barrier whatsoever to women for any type of vocation---and science has been richer as a result---I was nevertheless often amused that in legal cases, the attorneys in courtrooms would argue for admission to a male school or profession and adamantly profess that a woman could do anything a man could do, that there was no reason to exclude women from any job, while in the same breath insisting that the institution make physical changes in the work environment in order to accommodate women's Special Needs, a superb example of totalitarian doublethink. One other Stalinist aspect of this movement, which parallels the Lysenko affair, must be touched upon and this was that, in order to promote their ideology, the radical feminists and their sycophants (and why, why, *why,* does Stalinism, in all its variations, always attract so many sycophants?) carried out blatantly fraudulent research whose fraudulent results were widely and uncritically publicized by the mass media (Paglia,

1994; Sommers, 1994; Fox-Genovese, 1996; Pearson, 1997).

Second, we come to the question of intelligence and its measurement through IQ tests. Psychometricians have, for decades, through considerable research, attempted to objectively quantify intelligence through IQ tests and have found that, like physical traits, many psychological traits---like IQ---have a hereditary basis (Eysenck, 1973) which, if quantifiable, results in a bell-shaped curve. This, of course, has been condemned as being racist and Fascist and Nazi and so on and so on by individuals whose politics and overall outlook remarkably resemble the Stalinists.[10] They do not---ever!---offer experiments to contradict the above findings on IQ, but, instead, there is a lot of verbal abuse dished out. S. J. Gould (1981), in particular, wrote *The Mismeasure of Man*, one of the most contemptible works of deliberate, distorted, misrepresentation of scientific work; among other things, Gould uses an argument (for comparison's sake) that would condemn air travel today in airlines by pointing to the design flaws of aircraft in the 1800s and by citing the number of crashes occurring both from dirigibles and the primitive early airplanes. And aside from the question of IQ in the book, Gould also attacked Samuel Morton's work on skulls, claiming bias, even attributing physical statements on him which assertions have proven to be equally false (Michael, 1988; Lewis, et al., 2011). It has been my observation that, whenever someone rises up to defy the Stalinists in the West, to question their rigid dogma on any topic, to snicker at their self-righteous posturing, to question their unquestionable premises, the Stalinists mob him and overwhelm him with abuse and grossly misrepresent his position. They scream that he be dismissed from his job and his/her spineless employer, stunned from the sudden clamor and the viciousness, promptly caves in and the heretic is hurled into oblivion, followed by the thorough purging of that person's writings. To date, they have not been able to have anyone executed who disagreed with them or put forth ideological heresies---Western societies tend to frown on that sort of thing. Not that Stalinists would shrink from doing so, they can only dream of the day when such will be the reality. And if the offender happens to be already dead---such as Alexis Carrel, just to name one---then his name is stricken from monuments and various records, and he is assigned to oblivion. An unperson.

Third, we come to Edward Wilson's Sociobiology. Although he anachronistically calls himself a naturalist and is now diligently working on preserving biodiversity (Eisner, Lubchenko, Wilson, Wilcove & Bean, 1995), during the late 1970s he committed heresy in Harvard by publishing his work on Sociobiology which stated, in very brief form, that culture and human behavior had a biological basis (revealing his naiveté, he wrote (Wilson, 1994) that he could not understand why the staff and students and satellites in Harvard tried so hard to suppress the field of Sociobiology since Harvard had famously harbored people who had been accused of being Communists during the 1950s; he actually thought that the reason for Harvard embracing the accused Communists was because of Harvard's open-mindedness. How droll!). In 1978, as Wilson was participating in a symposium on

Sociobiology, as part of the AAAS, members of a Stalinist group, calling itself, typically, the International Committee Against Racism invaded the podium and bravely assaulted him (by the way, the man is blind in one eye and at the time had a cast on his leg), the first time in America that someone had been physically assaulted for expressing his scientific views. They then stood around telling each other how courageous they were. And Stephen Jay Gould, who happened to be present, apparently decided that this would be a good time to start spouting Lenin, which he did (Wilson, 1994) in his usual pompous manner. Nothing ever happened to the assaulters; considering that it was the 1970s, that is not surprising at all.

CONCLUSION

We have strayed far through a very convoluted path from the actual concept of inheritance through acquired traits, but I thought that it may be enlightening; I admit having always been both personally fascinated and repelled by fanaticism in all its many forms. Besides, we should always be on guard against fanatics invading the field of science and, if necessary, eject them with physical force. But aside from that, we have thus seen what happens *when a particular (political) theory becomes dogma to the point that it is not only not questioned nor tested, but facts become irrelevant, or ignored*, and if anyone points out contradictory facts, they are persecuted (we see this routinely taking place today in American and Western European universities by modern day Lysenkoists). Fraudulent research is also carried out to support such a theory. Even though the fraudulent research is eventually uncovered since, by the very nature of research it is public and verifiable, the damage is done.

We will now go on to the next, more famous theory.

FOOTNOTES TO CHAPTER 3

I use the word "scare" because, to be painfully honest, I can think of few things that would devastate my world view than a demonstrated need to return to the theory of evolution that is traditionally attributed to Lamarck.
---Richard Dawkins *The Extended Phenotype*

[1] Another, similar, case is that of Sigmund Freud. His ideas of personality and psychological development were the perverted ravings of a sick mind and many of his more acceptable ideas (e.g., the unconscious, the id) were old concepts, although he somehow made it seem to those outside the medical and the fledgling psychological fields as a conceptual breakthrough. Yet, he did come up with the unique and legitimate psychological concept of the different defense mechanisms (sublimation, denial, etc.) and of the famous "Freudian slip." Of a recent random survey of a dozen psychological textbooks that I made, only one mentioned Sigmund Freud, so it appears that he is finally, thankfully, fading from the scene, although editors in many professional journals tenaciously block any and all criticisms of The Master. Freud originated from Vienna, the same city that also gave us two other pseudosciences, mesmerism and phrenology. Incidentally, Freud was a physician, not a psychologist, although in reality he was an oneiromancer.

[2] It is for this reason that few good scientists have also been Marxists (or Nazis, for that matter). Most of

those scientists who have been Marxists and Nazis have been, almost always, again with a few exceptions, mediocrities in their field. Scientists seem to instinctively shy away from totalitarian ideologies, in contrast to intellectuals (journalists, lawyers, philosophers, writers, all whose job is to argue convincingly that black is white; Mussolini, Hitler, Lenin, Trotsky, Castro, Goebbels, Pol Pot, Saddam Hussein were all intellectuals). For further elucidation, I urge the reader to consult Eric Hoffer (1951).

[3]Yet another digression: during the early Thirties, there were two major prominent directors in Moscow, whose fame were worldwide, Meyerhold and Stanislavsky. The former employed a fantasy style, whereas the latter was all for realism. Their rivalry was finally settled in true Marxist fashion when Stanislavsky personally telephoned Stalin to complain, whereupon Meyerhold immediately became an unperson (Bowers, 1959). Stanislavsky's method of acting and directing are still much admired in the West. And speaking of the theater, George B. Shaw---who was Britain's second Lord Haw Haw during the Cold War---praised Lysenko and his triumph in 1948 over the geneticists (Huxley, 1949).

[4]When Medvedev's (1970) study on Lysenko was published in the West, he was placed in a psychiatric hospital by the Communist authorities (Bloch & Reddaway, 1977; Medvedev & Medvedev, 1971); the only amusing part to this macabre episode was when his twin brother showed up at the hospital director's office.

[5]It would be a mistake to think that it was just Lysenko and Prezent. They had many collaborators,

including one Olga Lepenshinskaya who, not only published fraudulent, pseudoscientific, research papers, but would regale her listeners with a tearful mien of the times when, having come to an impasse in her line of work, she would suddenly receive a telephone call from dear Comrade Stalin, who would then suggest a new approach to take in her investigations and which, invariably, yielded fruitful results.

[6]The disaster was not just a scientific and academic one. It had practical applications. Lysenko and his minions were put in charge of Soviet agriculture, which resulted, for decade after decade, in the Soviets being unable to feed their people and having to import food---this in spite of the fact that Russia had long ago subjugated the richly fertile Ukraine. Notice, by contrast, that the reverse occurred in physics, and the Russians were the first to put a satellite in orbit around the Earth, followed by being the first to put a man into space. Why? Because those "titanic geniuses" Stalin, Marx, Lenin, and Engels had never given any opinions on the subject on physics, atomic structure, or cosmology having never perused those fields---and if they had it would have been beyond their ability to comprehend.

[7]Richard Lewontin (of Harvard, of course): "There is nothing in Marx, Lenin, or Mao that is or can be in contradiction with the particular physical facts and processes of a particular set of phenomena in the objective world." (Wilson, 1994; p. 341) J. B. S. Haldane of Great Britain: "The commonest cause of gastritis---that is to say, an inflamed and irritable stomach---is worry and anxiety. It is particularly common among businessmen and traveling salesmen. I

had it for about fifteen years until I read Lenin and other writers, who showed me what was wrong with our society and how to cure it." (Zirkle, 1959; p. 488) Haldane also put forth the idea that the publication of Velikokvsky's book, *Worlds in Collision* was an American plot to prepare the population for a devastating nuclear war (Stove, 1976). Something that is particularly interesting is that during the first half of the Second World War, when Nazi Germany and Communist Russia were strong allies carving up Europe (a fact that is considered to be in bad taste whenever it is brought up among intellectuals), Haldane was a Soviet spy who was passing off military secrets to Russia, which in turn, passed on those secrets to Nazi Germany, so that the Luftwaffe could improve its bombing during the blitz; this revelation comes from the recently declassified Venona files (Romerstein & Breindel, 2000).

[8]The totalitarian implications of nature vs. nurture can be seen in Aldous Huxley's excellent novel, *Brave New World* and, to a lesser extent, in George Orwell's *1984*.

[9]One particularly tragic casualty of the female Stalinists is David Ramier. As an infant, he suffered a disfigurement as a result of a botched circumcision. Unfortunately, his parents made the mistake of consulting one Dr. John Money who had the bright idea of castrating Ramier and convinced his parents to henceforth treat the child as a girl in every respect, without deviation, and to administer female hormones as he grew up. Ramier was touted by Money as a success story that nurture was much more important than nature, i.e., environment over physiology and that, therefore,

since being a female was not a truly endogenous psychological state there were no endogenous limitations to being female. Money became very popular with the radical feminist movement, and he went on the lecture circuit, citing Ramier as proof. Meanwhile, things were going badly for the child. Ramier hated girl clothes and feminine activities, no matter how much they were crammed down his/her throat and gravitated towards masculine activity and dress. As Ramier grew older, he/she genuinely felt that he/she was going insane until he discovered the truth and tried to reverse the effects through surgery. He married, got divorced, and ultimately, committed suicide in 2004. Money, in the meantime, blamed Ramier's problems and sufferings on his mother, on his father, on television, on culture, on books, on Society, on everything and everyone except himself.

[10]Stephen Jay Gould was one of those individuals who attacked and deliberately misrepresented and ridiculed the concept of IQ (Orr, 2002) in a style that would have been the envy of any Creationist. He was so entrancing and charming when writing about paleontology and so repugnant, disgusting and unethical when doing so about psychology and politics. It is amusing, in an ironic sort of way, that someone who was so adamant in combating Creationism because Creationism did not want to accept a biological basis for human beings' bodies, also combated science because, as a Marxist, *he* could not accept the biological basis for human beings' behavior. Ethologists everywhere would be amused at that position.

Because he wrote the fraudulent work *The*

Mismeasurement of Man at the height of the anti-Arthur Jensen witch hunt, it was given awards by the National Book Critics Circle and the American Educational Research Association. The father of biodiversity, E. O. Wilson, referred to him as such, "I believe Gould was a charlatan. I believe that he was ... seeking reputation and credibility as a scientist and writer, and he did it consistently by distorting what other scientists were saying and devising arguments based upon that distortion." (French, 2011)

And speaking of ethologists, in 1973, when the Nobel Prize for Medicine and Physiology for Niko Tinbergen, Konrad Lorenz and Otto von Frisch was announced, European Stalinists agitated that the prize to the ethologists be rescinded because they felt that the recipients were all really a pack of Nazis. They still got the Nobel. By awarding the prize to ethologists in 1973, the fact was acknowledged that biology *is* the fundamental basis for human behavior, something that opponents to evolution, Sociobiology and the hereditability of IQ implicitly reject.

CHAPTER 4
THE FOURTH MECHANISM:
NATURAL SELECTION

Every species comes into existence coincident in time and space with a preexisting closely allied species.
 ---Alfred R. Wallace [The Sarawak Law]

On this same theory, it is evident that the fauna of any great period in the earth's history will be intermediate in general character between that which preceded and that which succeeded it.
---Charles Darwin, *The Origin of the Species Through Natural Selection*

When I open a page of Darwin, I immediately sense that I have been ushered into the presence of a great mind. When I read Phillip Johnson, I feel that I have been ushered into the presence of a lawyer.
 ---Richard Dawkins

What was it about Britannia (along with Germany and France) that in the 1700s and 1800s so many of its sons traveled to the four corners of the globe to explore, and, to collect and classify botanical and zoological specimens---sometimes attached to the Royal Navy and sometimes on their own initiative and often at their own expense? And this before the advent of true medicine, in areas so saturated with disease that, at times, they died in droves? (part of it may have been their acknowledgment---if Lyell and Darwin's writings are any indication---that humanity's actual knowledge of the natural world was practically nonexistent.)[1] As an indication of the *zeitgeist,* it is significant that from early on, scientists in Great Britain have been knighted, which in that country is considered a great honor (e.g., Sir Joseph Hooker, Sir Charles Lyell, both Darwin's contemporaries; curiously, neither Darwin nor Wallace were ever knighted). British civil servants, likewise, took advantage of their postings across the vast Empire to avail themselves of the opportunity to carry out scientific work, almost always thoroughly, even tediously. To take just two examples, by 1881, the only thorough knowledge of the natural history of Taiwan's avian and mammalian fauna was due to the British consul there, one Robert Swinhoe, who carried out extensive collecting and categorizing of specimens on his spare time and at his own expense (Wallace 1881/1998). In Java, when the British temporarily took over control of Java from the Dutch, Sir Thomas Raffles quickly seized the opportunity to excavate the mammoth Borobodur temple, a mound covered over ruin which had elicited no curiosity at all from the Dutch colonials. Indeed, some Indonesians have told me that if they had to have been colonized at all, they would have vastly preferred it had been by the British, a statement certain to send British Marxists into fits of apoplexy, who have consistently and deliberately distorted the history of the colonization period beyond recognition.

At any rate, the 1800s saw, on both sides of the Atlantic, a plethora of self-made scientists. We find one such naturalist, a self-taught, tall, shy man in a hut in the island of Ternate, in the Indonesian archipelago, in the grip of malaria. The man is Alfred Russel Wallace. During bouts of fever this Englishman thinks about all that he has seen of the fauna throughout his travels in the Dutch East Indies archipelago and in the Malaysian peninsula and what he has seen is very curious, most perplexing. He wonders about the various insects, mammals and birds and their relationship to each other and to their environment and their geographical distribution and in a sudden flash of insight he visualizes a new theory of what we would now call evolution. Within a week (actually, in three evenings) he has written a paper entitled, *On the Tendency of Varieties to Depart Indefinitely from the Original Type.* It is, in some ways, simply a logical extension of a previous paper of his, *On the Law Which has Regulated the Introduction of New Species,* which put forth what is nowadays referred to as the Sarawak Law, written when he had become immobilized because of a monsoon (Schilthuizen, 2004). Wallace had been deeply disappointed at the total lack of response to his Sarawak Law paper---with the exceptions of Charles Lyell, William Bates (his friend) and another naturalist by the name of Charles Darwin, who wrote Wallace to tell him that he agreed with every word written in the paper on the succession of species---since it was clearly the predecessor to the theory of evolution as we know it now (Shermer, 2002). Darwin had consoled Wallace at the lack of response to his previous paper (which Lyell had brought to his attention) by pointing out that naturalists were just interested in the description of new species and were blind to generalizations that could be made (Quammen, 1996).

The genesis of Wallace's theory of evolution can arguably be said to have taken root when he took a ferry from Java to Bali and Lombok (Wallace, 1998). Having spent considerable time in Malaysia, Java and Sumatra (seventy expeditions in an eight-year period!), he noticed that there was a distinct, unequivocal distribution of animals, particularly birds, all along the archipelago, even when the islands were very, very close. The eastern part of the archipelago had animals that were clearly related to those of Australia while the western part could be classified as of an Oriental variety; comparing species which were absent was just as impressive.

> In the Malay Archipelago [Indonesia] there are two islands, named Bali and Lombok, each about as large as Corsica, and separated by a strait only fifteen miles wide at its narrowest part. Yet these islands differ far more from each other in their birds and quadrupeds than do England and Japan (Wallace, 1881/1998; p. 4)

The biogeographical demarcation is now known as The Wallace Line and Wallace is considered to be the father of biogeography, perhaps unfairly, since earlier, in 1846, Salomon Müller had defined a similar line (mostly overlapping the later Wallace Line) based on ecology (van Oosterzee, 1997); Wallace (1890/2000) himself wrote later that a George Earl had earlier made the same observation and had published his findings in 1845; even so, as with Amerigo Vespucci, it is nonetheless nowadays called The Wallace Line. Incidentally, it has recently been noted that the line has been blurred as of late because humans tend to transport the animals of one region into another (Dunn, 2004); the truth, however, is that this contamination begun earlier (Wallace 1890/2000).

Contrary to his usual practice of submitting directly to journals, Wallace mails off a copy to an acquaintance of his in Great Britain, a well published naturalist with whom he has corresponded numerous times on various matters, by the name of Charles Darwin, requesting that, if the former thought it worthy, it be presented to Sir Charles Lyell, one of the contemporary giants of British science, but not to publish it. As luck would have it, Darwin had been sporadically working on a book whose thesis was the very same as Wallace's. In fact, both knew of each other's interest on the topic (Shermer, 2002).

This was not the first time that a scientific discovery was simultaneously made, and it would not be the last. A much earlier, famous instance was the discovery of calculus by both Newton and Leibniz which had led to bitter recriminations, as well as the subsequent discovery of oxygen by Lavoiser, Scheele and Priestley (Khun, 1962). In the 20th century, it almost occurred in the race to find the DNA molecule's structure. As someone has recently pointed out, this oddity only occurs in science; no two persons have ever simultaneously written a *Don Quixote de la Mancha*, or simultaneously painted *The Night Watch*. In fact, considering how intense the Newton/Leibniz debate lasted in Britain, I strongly believe that the subsequently Wallace/Darwin mini-controversy served as a catalyst to propel wide interest in the theory. Otherwise, it would have resulted in the same lack of interest as the Sarawak Law received.

At any rate, in 1858 Darwin was justifiably floored upon receiving Wallace's missive, "the Wallace thunderbolt," as Gale (1982) so dramatically puts it. It could not have come at a worse time: not only was Darwin very sick due to the chronic Chagra's Disease, but one of his children died days later and another child was deathly ill with diphtheria. Yet, Darwin and Wallace had corresponded with each other at length over the topic of variety within species, the Sarawak Law and the role of geographic isolation, especially islands (in his Sarawak paper, Wallace mentions the Galápagos Islands and the isolated animals therein). As late as 1857, Wallace had informed Darwin that he was still working on the questions raised in his Sarawak paper. Wallace took him by surprise with the speed that he finally came to the same conclusion as him; the reason for this is simple: Darwin was working on a book and obsessively searching for corroborating data, whereas Wallace was presenting a theoretical paper in a bare bones fashion. For his part, Wallace (and other scientists) knew that Darwin had been working on a book on species and variety of species for a long time (Gale, 1982; Shermer, 2002). It is not really as if Darwin had had no warning.

To be sure, for twenty years Darwin had been, on and off, ruminating over the idea and had been collecting such material as would support the theory (after all, Wallace's end product was a paper, whereas Darwin's was a book), trying to make it immune from criticism, but it is also widely acknowledged that he had been dawdling and procrastinating. Darwin had spent eight years working on a taxonomy of barnacles to the point that by the end he was sick to death of barnacles. Some modern scholars and scientists claim that that work, and others on various specific subjects, were a waste of time, a distraction, while other scholars argue that they gave him a good grounding in further scientific studies, which he himself had often acknowledged to have been deficient in. Incidentally, Darwin had originally become great friends with Lyell upon his return from the voyage (additionally, they had the same first names, both has abandoned their original studies in order to pursue scientific subjects, both were becoming prematurely bald, and, it would turn out that both of their theories would depend on a gradualist outlook) and Lyell, in turn, was both grateful for the data that Darwin had accumulated during his sea voyage that verified his theories (though they had disagreed on how coral reefs were formed, ultimately Lyell acknowledging that Darwin had been correct in his theory of coral formations) and had helped open doors in the scientific club for him, so much so that Darwin was very seriously thinking about pursuing his interests in geology, except for a sequence of events: first, he became ill with Chagra's Disease which prevented the mandatory field work of a geologist, second, he moved away from London where Lyell resided, so the daily visits ceased and, third, Darwin had by now married and his wife could not stand Lyell; as so many men have found out when they get married, their former good friends are subsequently excluded from the couple's company. Thus, geology faded into the background and into the foreground evolution slowly emerged.

  At any rate, Darwin contacted both Lyell and Sir Joseph Hooker, the famous botanist and a mutual friend, and informed them of what had transpired, in a prime example of British fair play which is one of those characteristics that make the British such an admirable people (and one can only smile at what would have happened if Wallace's letter had been sent to a Frenchman). Two weeks later, both proposals (Wallace's essay and Darwin's summary of his book) were read before the Linnean Society by proxy, and without Wallace's knowledge or permission, and both papers were published a month afterwards (in our cynical times, much has been made of Darwin's delay in notifying Hooker and Lyell of Wallace's missive, as well as on the odd absence of their earlier correspondence from Darwin's personal archives (Quammen, 2006)). The reader is free to interpret it any way that he wishes.

Wallace later wrote that his original essay had been a first draft, neither intended for publication nor presentation (Smith, 2003), but was unquestionably pleased, regardless, with the results which brought him instant fame and honor (as I have stated, I believe that his paper would have been ignored like his previous one on the Sarawak Law had it not been for (a) the sensation of the simultaneous discovery and (b) Darwin's book, which opened the theory to a wider audience). Upon his return to Great Britain, as a result of the presentation and publication, his professional standing was assured, and his finances improved to the point that he no longer had to travel and collect animals for collectors in order to make a living (as he had been doing up to that point). He subsequently became a prolific author on a kaleidoscopic range of topics, including, of course, evolution (Wallace, 1881/1998) and in 1889 wrote an entire book on evolution entitled *Darwinism* (Smith, 2004). At one point, Darwin was instrumental in getting a government pension for Wallace, who at the time was in need of financial security. Ultimately, Darwin was buried in Westminster Abbey while Wallace was buried in Dorset; in 1999, a subscription was launched to restore and protect Wallace's grave, which contains a seven-foot fossilized conifer trunk as a mysterious monument (Loder, 1999).

Charles Darwin, now no longer procrastinating, and no longer worried about offending the religious sensibilities of his friends and family, set about to finish his *magnum opus* once and for all. Previously, he had carried out meticulous scientific work on very specific, nontheoretical topics; a subsequent work of his establishes him as the first major ethologist (Darwin, 1872). A year after the presentation before the Linnaean Society, *The Origin of the Species Through Natural Selection* was published and sold out on the first day.

Pointing to Darwin's correspondence, many have pointed out that Darwin was rushed into publication of his theory by Wallace before he felt that he was truly ready, that is, before Darwin had conclusive, incontrovertible proof of evolution through Natural Selection. It sounds rather obvious and if Wallace had never written his paper, Darwin would have probably continued collecting material and developing his theory until the end of his days in spite of Hooker and Lyell's repeated urgings to publish, since Darwin's central argument, as has been often noted (Gould, 1995), is by analogy and not with direct evidence.

Darwin's *magnum opus* is important on several

levels. First, and most obvious, the establishment of the concept of evolution as a serious, respectable concept, of common descent. Long ago, Linnaeus had planted a seed in naturalists' mind with his taxonomy and the obvious conclusion was there that many plants and animals appeared to be related *and* that there was an increase in complexity of morphology. The idea of evolution was very definitely in the air and this must be understood because the Darwinian/Wallace theory was not formed in a vacuum as is so often thought: several vague theories had been bandied about for some time, all of them unsatisfactory (Secord, 1997); as was mentioned previously, Lamarck had proposed a theory, which had been subjected to skepticism; another Frenchman, a colleague of Lamarck named Etienne Geoffroy Saint-Hillaire put forth his own theory (Zimmer, 2001b); Darwin's own grandfather had previously suggested evolution:

> Organic Life beneath the shoreless waves
> Was born and nurs'd in Ocean's pearly caves;
> First forms minute, unseen by spheric glass,
> Move on the mud, or pierce the watery mass;
> Then as successive generations bloom,
> New powers acquire and larger limbs assume.

There was also an anonymous author, one Robert Chambers, who had published *Vestiges of Creation*, a work which was at the time considered to be rubbish, including by Darwin, even though it treated the topic of evolution, but, apparently, in a disreputable manner; T. H. Huxley, in a review, savaged it, which made Darwin nervous for his future work, if ever Huxley turned on it (Richards, 1992; Shermer, 2002). There is no doubt that the concept of evolution was there, floating around, but it had up to then been often presented in an absurd manner, or, as a vague concept.

There is also a relevant psychological element involved and that is the obsession that Britain had at the time with *progress,* based on the endless discoveries and inventions taking place all the time. The very essence of evolution implies *progress,* an increasingly complex and sophistication of organisms. The concept of progress was in the air (Carr, 1961). Darwin himself noted that.

Second, Wallace and Darwin put forth a rational, logical, convincing *mechanism* by which evolution comes about, viz., Natural Selection, whereby individuals possessing advantageous traits do better in the business of surviving, have more offsprings as a result, and a steady accumulation of such traits ultimately lead to new species, different and more successful than the old.

Third, Darwin, like Lamarck, emphasized the importance of variation *within* a species. People were aware that there were differences between members of a species, but, aside from breeders, their true significance was pretty much ignored by most natural scientists. The prevalent view was that there was a Platonic ideal of each organism's morphology, from which unfortunate variations occurred (Greek philosophy was still firmly embedded in European intellectual thinking).

Fourth, like Wallace and Lyell and many others, he pointed out that there was a succession of species, many of which had become extinct. The fact that previous animals, and monsters, had existed before and had become extinct, had been known for a long time thanks to a large part to the work of Georges Cuvier, not to mention the fossils that people at times seemed to trip over; indeed, by the time of the Darwin-Wallace publication countless antediluvian creatures were being frantically dug up on both sides of the Atlantic (Jaffe, 2000). In fact, just a few years after the publication of *The Origin of the Species*, T. H. Huxley exulted over the discovery of the famous *Archaeopteryx* fossil in the Solnhofen two years after Darwin's publication as a supposedly fortuitous piece of crucial evidence of a missing link, this time being between dinosaurs and birds (he later was likewise grateful for the horse's ancestors being discovered, which served as a cornerstone of one of his famous lectures).

Fifth, he brought together a huge collection of data and information, which had hitherto been unrelated and which he arranged around his theory. We take this interrelationship of data for granted nowadays, but at the time even the word "biology" did not exist and what naturalists did was more in the way of endless classification and description than anything else: this bird has this plumage, this monkey has this type of diet, the skeletal structure of the bison follows similar pattern to that of the elk, this ant lives here, this animal hibernates there, etc. It could be said, rightfully, that the theory gave purpose to biology, gave it a central goal, gave it life.

Sixth, without saying so outright, it became obvious that religion played no role at all in biology and that there was no magic involved. Most books on Nature, even those written by scientists, invoked The Creator in their description of animals and plants, particularly exotic ones and some even dripped with sentimentality. Ernst Mayr (2001) put it very well: "No wonder the Origin caused such turmoil. It almost single-handedly effected the secularization of science." (p. 9)

Seventh, it emphasized the intraspecies competition, whereas most persons were aware of interspecies conflict.

(One of the irritating aspects of scholars is their habit of, once an individual has made his mark in the history of science, to go back in the records and search for any and all precursors and predecessors, thereby intimating that their contribution was much less than what it is attributed. So, it is with Lamarck (Zirkle, 1946) and so it is with Darwin (Zirkle, 1959; Greene, 1961, 1963; Richards, 1992; Gale, 1982)

For all of the classical theory's virtues, there is one inescapable fact: there is no direct, conclusive evidence of Natural Selection as the force behind evolution. Nor is it testable. It is an exceedingly reasonable hypothesis, supported through the analogy of selective breeding by humans of livestock, plants and pets and through circumstantial evidence. Nonetheless, Natural Selection has never been proven, it has always been inferred.

And it is important to separate the concept of evolution from the *mechanism* of Natural Selection, which is one of the central points in this book. This semantic difference has tripped many a scholar and scientist.

* * * * * * * *
* * * *

As a young man, Charles Darwin had begun and abandoned several professions, much to his father's exasperation. He had shown a natural inclination towards natural science; his grandfather, Erasmus Darwin, had been a famous naturalist and had himself hinted at the concept of evolution; a friend of his at college had been an enthusiastic proponent of Lamarck (it has been forgotten in what truly dismal state was the scientific teaching in British universities, unlike German universities, to the point that in 1845, there were practically no scientific laboratories in universities (Nature, 1998)). In 1831, then, dissatisfied with his schooling, he joined an exploratory and scientific expedition aboard the *H.M.S. Beagle* as an unpaid naturalist. During the two year voyage, which turned into five years, Darwin read Lyell's *Principles of Geology: Being an Attempt to Explain the Former Changes of the Earth's Surface by References to Causes Now in Operation*, wherein Lyell argued for Earth's antiquity and a gradualist outlook in explaining geological formations (as opposed to catastrophic explanations) and which was an eye opener for both Darwin and Wallace among others (up to that time, it was generally thought by many that the world had been created circa 4004 B.C.). As an eminent historian has pointed out, Lyell brought history into science (Carr, 1961). Curiously,

Lyell devoted eleven chapters in the second volume of his work on geology on, instead, biology, four of which were a critique of Lamarck (one wag has stated that Lyell's book is actually *The Origin of the Species* without the presence of Natural Selection (Allen, 1994); it certainly must have planted a seed or two in Darwin's head). Many years later, Lyell would admit that his aversion towards Lamarck was the latter's claim that man had descended from the apes (Secord, 1997).

Lyell's work, though unfairly overcriticized recently by Benton (2003), is interesting reading nowadays if for no other reason than it describes the slow formation of the science of geology, a science which was affected by the politics of the time. One passage is relevant, however, to our overall present discussion (Lyell, 1833/1997, p.11), for which I will momentarily digress:

> Unfortunately, the limited district examined by the Saxon professor was no type of the world, not even of Europe; and, what was still more deplorable, when the ingenuity of his scholars had tortured the phenomena of distant countries, and even of another hemisphere, into conformity with his theoretical standard, it was discovered that 'the master' has misinterpreted many of the appearances in the immediate neighbourhood of Freyberg.

At any rate, as the *Beagle* traveled through South America, Darwin enthusiastically collected specimens of plants and animals, excavated fossils and examined geological formations. It was not until the ship reached the Galápagos islands, where it spent five weeks, that the naturalist's curiosity---and puzzlement---may have peaked, although his documentation of the specimens that he collected in those islands was sloppy (Darwin, 1860; Moore, 1964). What ultimately intrigued him was, first, that the animal species in these islands were endemic to the islands even though they were unquestionably related to other species in the mainland, second, that the inhabitants of each island differed from those of the other islands (the vice governor of the island noted that he could always tell from which island originated any tortoise presented to him) and, third, that----particularly in the case of finches---species had filled every ecological niche in each island due to some unique advantage that each species had over the other. The Galápagos have been dubbed both as Darwin's Laboratory and Nature's Laboratory, but it surprises me that they have not been seen as an example of the scientific method, to wit, inferring from the population being studied through sampling; to Darwin, the Galápagos turned out to be a sample, a sample of biological diversity from which to infer general principles from overall Nature. Incidentally, at this time Darwin saw himself primarily as a geologist, due to Lyell's influence (his influence would also become evident in Darwin's emphasis on *gradual* evolutionary change).

The *Beagle* returned to the UK in 1836. Darwin had, from time to time, mailed ahead most of the specimens that he had collected.[2] Henslow, his mentor from his college days, had read his periodic letters before the Cambridge Philosophical Society and the Geological Society of London. Charles' father, upon being told that his son was doing excellent work, which foreshadowed his becoming a great man, was astonished (Marks, 1991). For the next two decades he ruminated, he re-examined his collection and he conducted numerous experiments that would buttress his observations. During this time, as stated previously, he published works on very specific topics.

*   *   *   *   *   *   *   *
  *   *   *   *

I have just finished rereading *The Origin of the Species*. It confirms the impression that I got on my first reading it: it is a wonderful, wonderful book, which with its felicitous language is a pleasure to read. One thing that I find particularly endearing is his admission when he states (and I am paraphrasing here), "Look, this is the theory that I am putting forth. I admit that at present we do not have all the facts yet, especially in the fossil record, which is spotty, but I have supreme confidence in future investigators and that they will fill in the gaps and the theory will be corroborated." [3] And, of course, he was right. Gregor Mendel, an Austrian monk and, ironically, a contemporary of both Wallace and Darwin, had just then carried out botanical experiments which went a long way in describing the mechanism of heredity (but more about this later), but whose results remained in obscurity.

The Darwin-Wallace theory of evolution, which I will refer to hereafter as the classical theory, is simplicity itself and in its simplicity echoes the state of science during the nineteenth century. It is so simple that the theory can be stated and elucidated in just a few pages---

which is what Wallace had done. The rest of the book by Darwin is simply elaboration, padding, and restatement.

At the risk of repetition, let me restate it: the theory points to the obvious fact that individuals within a species will vary from one another in any number of ways. Additionally, the amount of food, water, mates, nesting materials and spots are always limited, so that there is fierce competition for those limited resources, not only with conspecifics but with other species. In this, both men Darwin and Wallace had been greatly impressed by Malthus' 1798 *An Essay on the Principle of Evolution*[4] (as well as Lyell's work, it must be remembered); indeed, Darwin wrote that he now had a peg on which to hang his ideas. Consequently, any individual which possesses a trait that will give it an edge over its competitors will, as a result, have more offsprings and they, in turn, will have that trait. As environmental circumstances change, any other traits that give an individual an adaptive advantage over its competitors will have similar results. These cumulative traits, through successive generations, will ultimately result in a new species, different from the original one (essentially, this turned Lamarck's argument around: those giraffes with long necks were better adapted, they multiplied and produced more descendants rather than giraffes who strained their necks produced offsprings with longer necks). This process, this mechanism, Darwin called Natural Selection (and the phrase "survival of the fittest" was coined later by Herbert Spencer to illustrate the process; the word "evolution" was used only once by Darwin in his work, but Spencer kept it alive). In so far as Natural Selection is a constant, ongoing process, new species are always in the making. As mentioned in a previous chapter, some scientists have gone so far as to seriously put forth the proposition that there should be no species nomenclature at all because all species are in a constant state of transition (Gould, 1982), an extreme view, to be sure. And, just as importantly, because the culmination of the process into a new species takes such an exceedingly long time, we cannot detect it.

Furthermore, Darwin pointed out at great length that although Natural Selection is a process that takes a very, very, very long time to "complete," human beings, through artificial manipulation, i.e., selective breeding, can greatly accelerate the process. Thusly, in the case of plants, we have an almost infinite variety of roses, orchids and tulips for which humans have selected for (this must have especially struck home with the British public). The same is true for crops which are selected for a greater, or tastier, yield.[5] The very same procedure is carried out with animals, either livestock, or pets. Siamese Fighting Fish (*Betta splendens*), for example, have been selectively bred by breeders not only for more aggressiveness, but for appearance, so that the specimens offered up for sale in pet store have luxuriant growths of fins with spectacular colors, compared to which the wild variety is a drab comparison. Even more interesting, domestically bred species of *Bettas* linger near the surface to gulp air, whereas the wild variety avoids predators by darting to the surface followed by a quick dive (Wolfsheimer, 1975).

The evolutionary theory put forth was, at the time, the most coherent one, supported by much evidence, thanks to Darwin, one that contradicted the idea of the immutability of species, as well as the idea of individual creation of each individual species. The Wallace-Darwin theory instantly gained favor with most scientists partly because of the rational, logical style with which the theory was presented and partly because of the mechanism proposed that accounted for the transformation of species (i.e., Natural Selection) which appeared to be logically supported by the data. And it also became popular with the public because it made them aware, for the first time, of the concept of evolution. It is really remarkable how quickly the theory became disseminated among both scientists and the public.

Nonetheless, it immediately came into unremitting attack; most of the acrimonious attacks on the theory, though, were based on religious grounds. I have often wondered why. After all, Lyell's work had flatly contradicted Genesis as the earth having been created in six days; he made it clear that the Earth's present geological appearance had been formed through a vast period of time; this conceptualizing of such vast time was a novel, revolutionary, one which we now take for granted (and which planted a seed in Wallace and Darwin's minds). Would the attacks have come if only Wallace's paper and not Darwin's book had been published (for that matter, would anyone have noticed?)? Would the attacks have taken place if the famous debate, with its mutual verbal jibes, between T. H. Huxley and the Bishop of Oxford had not taken place? Was it perhaps because of a sentence[6] in the first edition that Darwin had written that the same process applied to man: "Much light will be thrown on the origin of man and his history"?[7] It became apparent that rather than the previous popular view that Man had been placed on this Earth, Man had, instead, been part of Nature from the very beginning.

To be sure, there also was some criticism of the theory on scientific grounds. Ironically, they came from Darwin's main supporters, Hooker, Huxley, Lyell and even Wallace. Some of them were of the opinion that Natural Selection could not possibly be the only force in causing evolution. Huxley, in particular, bewailed the lack of experimental data to corroborate the theory (it is important to point out that Darwin did subsequently perform experiments that peripherally supported his thesis, such as (a) suspending seeds in seawater for days, or weeks, in order to ascertain whether they germinated and thus colonize distant islands, some of which germinated after one hundred days' immersion (b) planting untended gardens with many different species of plants and observing that there was, indeed, a struggle for existence between plants until only a few remained of the original mix; undoubtedly, he got the idea of this particular experiment from Decandolle, who is mentioned in Lyell's work). Perhaps as a result of these criticisms (or perhaps he meant to do so all along in his original idea of a much larger book) Darwin later wrote a work elucidating Sexual Selection, whereby a trait is selected for, not because it is beneficial in coping with the environment, or in obtaining food, or avoiding predation, but because its mate finds it appealing (e.g., the peacock's tail feathers). Ironically, he also introduced his own Lamarckian-type theory of "pangenesis" with which to buttress his theory, but pangenesis was just rubbish; to my knowledge, pangenesis is the only major error in his prolific writings. Or, as is much more likely, perhaps Darwin had intended all along to publish both his works united as one, but Wallace's more narrowly focused paper forced Darwin to publish an "abridged" version.

The last years of his life were spent, not just on revisions of previous works, but also on important new works. Honors came his way from European countries, with the exception of France (Gillman, 1996)

*   *   *   *   *   *   *   *
*   *   *   *

While teaching in colleges, I at one point in the semester used to ask my students to state all the similarities between men and women. Invariably, there was silence, and confusion could be read in their faces and it was some time before one or two would venture a guess. They, and everyone else, for that matter, was, and is, used to constantly, obsessively, looking at the differences between men and women. The same is true for human beings towards animals and for centuries philosophers, scientists and laymen have debated on what sets Man apart from "the dumb beasts;" some have said the soul, others the size of his brain, still others the opposable thumb and/or tool using ability (I, for one, feel that what makes Man truly unique and wonderful is his ability to create atomic weapons with which to completely obliterate one another).

When scientists began to compare the amount of

DNA in chimpanzees that was similar to human DNA, other scientists began placing bets with one another. Some said that the degree of genetic similarity would reach 30%, others around 40%, while some brave souls predicted as much as a 50% genetic similarity. As work slowly progressed, the degree of similarity passed 30, then 40, and then 50%, and you could actually begin to hear nervous laughter among scientists. Finally, when it reached 98.5%, there was a stunned silence. But scientists, of all people, should not have been surprised at this finding. Chimpanzees have been used in medical research because what affects them medically affects us as well; if this was not the case, chimpanzees would not be used and they could breathe easier. But, most importantly, it is the fact that when one physiologically examines an ape, we see the same biochemical processes that are so precise, so delicate, so finely tuned, so unbelievably complicated in their function, with each step having so many other ramifications. You have the endocrine system, the reproductive system, the circulatory system, the nervous system, the eyes, the liver, the kidneys, the blood, homeostasis, the lymph glands, the brain, the reproductive system, each and every one of these entities so breathtakingly, so mind bogglingly complex that each one would fill all the volumes of the Encyclopedia Britannica to describe each system in detail (and a breakdown in anyone of those countless details can result in death for the organism). There are certain physical disorders that are caused by a single misplaced "letter" in a single gene (Hall, 1990). A number of genetic diseases are known where a single amino acid change destroys hemoglobin's ability to carry oxygen effectively (Behe, 2007). This can be compared to having the entire *Encyclopedia Britannica* voided because one word is misspelled (e.g., "the process whereby a protein sequence is produced, from the triplet code in messenger RNA, turned out to be extraordinarily complex." (Steele, Lindley & Blanden, 1998; p.42)).

Nor are chimpanzees the only organisms that we use for medical research. We use rats, pigs, rhesus monkeys, dogs, guinea pigs, according to an institution's finances and whether one of the other organisms happens to have a physiological uniqueness that excludes it from that particular research avenue.

It must also be remembered that Fritz Lipmann of Rockefeller University replaced transfer RNA from rabbit hemoglobin with transfer RNA from *E. coli* without destroying the normal process of protein formation (Moore, 1964). Indeed, the basic structure of several genes can be traced for millions of years, a truly astonishing fact, not to mention the molecular clock (Mayr, 2001).

> Virtually no biologist expected to find what turned out to be the case: most of the genes first identified as body-building and organ-forming genes in the fruit fly have exact counterparts, performing similar jobs, in most mammals, including humans.

(Carroll, 2005, p. 59)

Indeed, to take one example, the *Hox* genes, which dictate the head to tail morphology in fruit flies (and which are curiously aligned to the animals' morphology), have their counterparts in those of mice or geese (Zimmer, 2006). Plants have a corresponding MADS-box genes (Carroll, 2000).

That evolution has, indeed, taken place through time becomes obvious if we look closely at both plants and animals, apart. All plants are closely related to one another and all animals (at least as far as vertebrates are concerned) are closely related to each other. But it is the similarities, or the evolution if you prefer, of plants that is so much more obvious than in animals (except for a botanist, or a gardener, or a horticulturalist, all trees look pretty much alike to an onlooker). Since trees look so much alike to the nonbotanist, and, truly, the differences between species of trees is minimal, compared to animals, the concept of evolution is easier to teach in children if one uses plants instead of animals.

Take, for instance, a human, a gorilla, a cheetah, a tapir, a moose, a bat, a frog, a macaw and, yes, even a *T. rex* and a *Stegosaurus*. All of these animals are practically the same. They are almost identical to each other. External morphology is very distracting, yet it is only skin deep. Even apart from their internal biochemical processes, they are variations of the same animal: all of them have two arms and two legs, a tail, a head and a rib cage surrounding the upper thorax. True, there are minor variations with each other, to be sure, like the human has a vestigial tail (the coccyx) while the *Stegosaurus* has four spikes in its tail and those plates on its back, but these are minor, external differences. No vertebrate animal on earth has six or eight limbs, or a rib cage that extends to protect its underbelly, nor two hearts, or four kidneys---even though they would all be adaptive in the extreme. And, they are all, roughly, cylindrical in shape.[8] And it is no accident that humans and many animals are subject to the very same diseases (e.g., chimpanzees and polio, dogs and rabies, armadillos and leprosy). It is also no accident that humans and plants are *not* subject to the same illnesses. An additional point is that archaeologists and paleontologists are often able to reconstruct a whole animal or person from a mere half dozen surviving bones out of the entire, vanished, skeleton---and sometimes from just a molar.

And this brings me to point out a cardinal rule of Nature, what I call Nature's Paradox: *infinite variation through constant duplication*. It is as if Mother Nature (I know: I am anthropomorphizing, but it is simply to make a point) comes up with an idea for an organism and just will not let go of that idea. Look at all the different species of extinct trilobites. Look at all the sauropods. Or, take living examples: think of all the thousands of species of starfish, or birds, or spiders, or, better yet, horned herbivores. Look at trees or at beetles.

Nor does it have to be an entire organism; it can be a component of an organism. For example: the beak found in parrots can be found in the extinct *Psittacosaurs* like the *Triceratops* and *Protoceratops* and it can also be found in, of all things, squids. This could be cited as an example of convergent evolution. But, is it? More about this later.[9]

In short, the concept of evolution is a given. It is the *mechanism* for bringing about speciation, i.e., evolution, that we will be calling into question.

FOOTNOTES FOR CHAPTER FOUR

If we compare corresponding portions of different continents, we find no indication that the almost perfect similarity of climate and general conditions has any tendency to produce any similarity in the animal world.
---Alfred Russel Wallace, *Island Life*

[1]Soon after finishing this manuscript, I came across a fascinating book that deals precisely with this question, *The Naturalist in Britain* (Allen, 1994), the only one to my knowledge to do so. Wootton's (2016) superb *The Invention of Science* also addresses this.

[2]Darwin's collection arrived but Wallace's did not. In 1852, Wallace had finished his travels in the Amazon and the boat in which he was sailing burned and his collection either burned or sank. There must have been an evil water deity with a grudge against the British in those days. Josef Conrad's first manuscript suffered the same fate as, to a certain extent (it was in a train this time), did Lawrence of Arabia's work on history.

[3]In this he reminds me of America's Founding Fathers when they put forth their Constitution after much wrangling; they knew, and said so, that it was imperfect, but had confidence in future generations and were confident that those future generations would get the kinks out. And, of course, they were right.

At any rate, it appears evident that nineteenth century scientists, unlike today's scientists were humbly aware of just how little was actually, reliably known of the natural world. Lyell (1833/1997) himself said just as much (p.76):

> If this remarkable break in the regular sequence of physical events is merely apparent, arising from the present imperfect state of our knowledge, it nevertheless serves to set in a clearer point of view the intimate connexion between great changes in the physical geography of the earth, and revolutions in the mean temperature of air water.

[4]Which thesis, by the way, found confirmation in the horrific 1845 Irish Potato Famine (Gray, 1995).

[5]After reading Darwin's huge volume, Wallace was impressed and wrote, "Never have such vast masses of widely scattered and hitherto utterly disconnected facts been combined into a system and brought to bear upon the establishment of such a grand and new and simple philosophy." (van Oosterzee, 1997; p. 159).

[6]There have been other instances in science where only one or two sentences written in a body of work have created a sensation.

[7]A curiosity of the religious attacks on evolution is that the most persistent attacks on religious grounds has been in Anglo-Saxon countries. One can only speculate as to whether this strife would not have become nonexistent if Huxley (the self-appointed "Darwin's bulldog") had not purposely returned Bishop Wilberforce's insult. To my knowledge, there has been no comparable continual attacks on the theory in Catholic countries; it is a non-issue. Additionally, the Catholic Church genuinely repented, and learned from, its treatment of Galileo. One has to remember that the Catholic Church has always had an intellectual, scholarly

foundation, unlike many contemporary Protestant sects whose whole foundation seems to be hysteria. Indeed, I have repeatedly found Pentecostal and Southern Baptist ministers in the United States to be complete idiots. The writings in the books of these denominations are pure drivel; like the speeches of Billy Graham, they go on and on and on while saying absolutely nothing. You will not find in them the equivalent of a St. Thomas Aquinas, a St. Augustine, Tielhard de Chardin, or Thomas a Kempis. Can the reader think of even one fundamentalist Protestant minister who has also become a scientist, someone of the caliber of Tielhard de Chardin? I cannot. And what is curious is that of the Protestant branch, most of the ministers who have also been naturalists or scientists have been of the Anglican Church, which is the closest to the original Catholic Church, although they would vehemently deny it.

One exception to the religious attacks on the classical theory has been the Mormons in that Brigham Young welcomed the science of paleontology because it confirmed one of the minor points in the Mormon religion, which is that there were horses in the Americas prior to the Spaniards introducing them in the 16th century (Jaffe, 2000).

As for Islam, I regret to say that I have no first hand information on how that religion has reacted, systematically, to the theory of evolution, other than the occasional fanatic (Quammen, 2004), and in 2017 the Turkish government stopped the teaching of evolution in schools (Mika, 2017). I strongly suspect that the degree of fanaticism in regards to the subject of evolution varies from Muslim country to Muslim country, with the theocratic countries of Iran and Saudi Arabia (and, until recently, Afghanistan) dealing with scientists who adhere to the theory in the usual way that they deal with people who dance, sing, wear fingernail polish and who drink a Cuba Libre: torture and death. However, in Indonesia, a multiethnic archipelago-country, the teaching of evolution is not prohibited, aided partly, no doubt, to national pride, since Wallace did form his theory while there, in one of the islands. Indeed, the city of Bandung has streets named after famous scientists, including one of Wallace and one of Darwin.

Having said that, let us see an instance where in a Muslim country, science meets religion head on (Sikorski, 1990):

> President Zia Ul-Haq himself was to inaugurate the conference in four months. They spoke in faultless, somewhat fussily accented English.
> 'Most interesting paper arrived yesterday, by facsimile from Jeddah'---the lean one stressed the 'by facsimile.' 'This one will make quite a stir, mark my words,' he continued. 'This is strictly off the record, you understand?' He looked at me sharply. 'On the basis of the Koran'---he lowered his voice 'Professor Nuri of the Islamic University of Saudi Arabia has made a key discovery in physics, which will be officially revealed at the conference.'
> 'What is it?' the chubby one asked---out of politeness, I felt.
> 'Well, I think it's sensational. Einstein is transcended. The speed of light in the universe cannot be limited, because if it was it would take an angel descending from the nearest star four years to reach us, whereas Gabriel used to descend

> to the Prophet, peace be unto him, several times a day. So simple, isn't it? Mind you, the President is particularly interested in a paper on generating electricity from djinn. The djinn'---he turned to me again---'are creatures which the Koran says are made up entirely out of fire. If only we can find a way of luring them to one place in large quantities . . . you understand? That would at last stop the Americans from complaining about our nuclear energy programme. It would satisfy thirty percent . . . .' (p. 19)

[8] In 1979, I contributed a chapter to a book, *UFO Phenomena and the Behavioral Scientist*, edited by Richard Haines, and in it I argued that since the reports of flying saucer occupants were always described as humanoid, they could be discounted, in as much as it was extremely improbable that the evolution of two different planets would end up with the same end products at the same time with the same motivation. More credibility could be attributed had the occupants been described as "monsters." Additionally, the possibility that those extraterrestrial species had become "intelligent" *and* had also developed the attributes of curiosity *and* exploration *and* were as technologically advanced, or more so, as us *and* in the same types of technology was out of the question. I was amused on learning, in 2004, that Alfred Wallace had made the same point back in 1904 (Smith, 2004).

And while we are on the subject, please note that on this planet there are primarily two types of organisms which are totally alien to each other: plants and animals.

[9] We tend to extol the great biodiversity found in

Nature and bewail its diminution (Tilman, 1996; Leach & Givnish, 1996; Wahlberg, Moilanen & Hanski, 1996; Perlman & Adelson, 1997; Terborgh, 1992; Moser, 1975;Crooks & Soule, 1999; Schoener & Spiller, 1996; Spiller & Schoener, 1990; Brooks & Balmford, 1996; Kideys, 2002; Verissimo, Cochrane & Souza, 2002; Allen, Brown & Gillooly, 2002; Curlotta, 1996; Saether, 1999; Morell, 1996; Moore, 1999). It has often been said that life is wonderfully varied, with an almost infinite number of animals. I see the reverse. In actuality, what we constantly see in Nature are endless variations of a very few basic themes. There is a dearth of original patterns, just a handful really, but an almost infinite variation upon the basic structures.

CHAPTER 5
THE FIFTH MECHANISM: MUTATIONS

Only those who remember the utter darkness before the
Mendelian dawn can appreciate what happened.
---William Bateson

Mutations, in summary, tend to induce sickness, death, or deficiencies. No evidence in the vast literature of heredity change shows unambiguous evidence that random mutation itself, even with geographical isolation of populations, leads to speciation.

---Margulis & Sagan

I spent an especially memorable evening arguing with [Isidor] Rabi about the evolutionary consequences of atomic bomb tests; he defended the position that the explosions were good because radiation increases the rate of mutation, which can speed up evolution. And that is a good thing, is it not? Was he serious? I was not completely sure.
---Edward O. Wilson, *Naturalist*

The Darwinian-Wallace theory was not accepted *in toto* by a number of scientific skeptics and their skepticism, in a way, is really a reflection of the dismal state of science at the time.[1] It was not until 1831, for example, that Robert Brown discovered the cell nucleus (and by doing so indirectly led to a mania for microscopes and their improvement (Allen, 1994)). Nor, in 1837 that Schwann and Schleiden realized that the cell was the basic unit of life, or in 1855, that Rudolph Virchow established the fact that cells within an organism's body formed through cell division, or in 1865 that Schweigger-Seldel and LaVallette St. George showed that spermatozoa were cells, or in 1875 that Oscar Hertwig demonstrated that fertilization of an ovum was due to a single sperm, while in 1883 William Roux hypothesized that the source of heredity resided in the chromosomes. So one can see that bits of knowledge that nowadays is so taken for granted that grade school children know of them, were, as late as the mid-1800s, revolutionary, and sometimes controversial, discoveries.

Switch now to Brunn, Austria. A pleasant Austrian monk by the name of Gregor Mendel, who had always been fascinated by Nature, carried out an exhaustive experiment in the crossbreeding of garden peas. On reviewing the literature, Mendel had been struck at how chaotic the previous studies had been conducted (it has not been properly acknowledged that one of the factors for the rise of knowledge and the rise in the overall standard of living in the 1800s was greatly due to the realization of the concept---and the application---of "methodical"). In 1865 he read the results of his experiments before the Brunn Society for the Study of Natural Science. It went right over the head of his listeners. The audience almost certainly were perplexed at the linking of botanical observations with mathematics, even though the mathematics that he introduced would be considered nowadays to be elementary, even simplistic, statistics (Barber, 1960). Nonetheless, his monograph was printed up and distributed, as was customary, to various scientific institutions (more than 120, to be exact (Rhoades, 1992), where it lay, unread, a scientific ticking time bomb.

It should be noted that Mendel did his work independent of the evolution controversies.[2] Although his findings became well known throughout Europe after Darwin's death, Wallace was still alive and still very prolific in his writings, wherein he cited Mendel's work.

Switch again to Amsterdam, Holland. Hugo De Vries worked for years with the evening primrose on a series of botanical experiments which were as extensive, if not more so, than Mendel's. Then, in 1900, on the threshold of announcing his findings on numerical ratios of hereditary units, he came across Mendel's work in a reference of a book that he was consulting. Exactly a month later, a German scientist, Karl Correns, who had been working with maize and peas, came forth and reported the same ratios and he, too, had just discovered Mendel's monograph. And then, exactly another month later, another Austrian scientist, Erik Tschermak, who had been working with peas, reported the very same thing. All four had proven that many hereditary traits were particular units, not global, not a Gestalt, not a blending, and, furthermore, that the heredity of those particular traits was in a discernible ratio.

De Vries, along with many other scientists, had been dissatisfied with the mechanism of Natural Selection as being able to explain speciation. He saw that human selective breeders of animals and plants could only select for those differences that were already in existence; otherwise, they had to wait for radical changes in Nature which De Vries called "mutations." To him, it was the mutations that were the main mechanism for evolution. Once the mutation occurred, Natural Selection went into action only to determine whether it was adaptive or maladaptive. He saw evolution as a series of sudden jumps, rather than a tedious, gradual accumulation of minor traits. Darwin had predicated Natural Selection on variations that already existed in a species. He did not foresee that new traits (variations) occurring through "mutations," upon which Natural Selection would act on. In other words, the classical theory stated that evolution occurs about because of traits that are already present in the organism which for some reason, now become beneficial-adaptational, whereas with the theory of mutations, brand new traits suddenly occur which result in new species.

The mutationists concentrated on Batesian mimicry in their argument. According to them, the physical mimicry of one insect by another insect could not possibly have taken place by Natural Selection, it had to be the work of mutations, i.e., Natural Selection could destroy, but not create species.

An American mutationist, T. H. Morgan, working with what was to become the geneticists' workhorse, the fruit flies, found that certain traits seemed to be linked, which, in turn, meant "linked genes," the most obvious instance being sex linked traits as with color blindness in men (by the way, Morgan's scientific work was to earn him undying hatred amongst the Communists).[3]

A Briton went to the defense of his deceased countryman. Sir Ronald Fisher simply overwhelmed the mutationists with mathematical equations, proving that it was mathematically impossible for the same pattern of mimicry to have taken place through independent mutations. Apparently he thought, if he did not say so outright, that the individual development of the eye in 40 different branches of the evolutionary tree was mathematically possible.

At roughly about the same time, attempts were being made to induce speciation in *Drosophila* by applying heat, cold, poison, drugs and mutilation. H. J. Muller, who had worked with Morgan, then used X-rays with success. The eyes, wings and bristles of the fruit flies were altered, often radically, as a result of the X-ray bombardment (Srb, 1965; Gardner, 1975). It should be pointed out that all of these mutations were distortions; at no point was there any speciation. No fruit flies were turned into horseflies, butterflies, or beetles.

It is now taken for granted that "mutation pressure" alone cannot cause evolution.

Incidentally, the atomic bombs dropped over Nagasaki and Hiroshima did not produce any speciation in humans, animals, or plants. Animals and humans that survived the initial effects of the blast for years thereafter suffered various illnesses associated with radiation poisoning, some of the illnesses being terminal, and could not find marriage partners for the inhabitants outside of those cities, even those born long after the explosions.[4] Nor in Chernobyl decades later. Artificially induced mutations in complex, higher, organisms almost always have deleterious effects on the organisms to the point that they can be lethal.

Ernst Mayr (2001) went step further. Speaking of naturally occurring mutations, Mayr actually stated that the reason that we do not see beneficial mutations cropping up in nature is because they have all been used up:

> Individuals with genotypes that contain a beneficial new mutation will be favored by natural selection. However, since almost all conceivable beneficial mutations of a population in a stable environment have already been selected in the recent past, the occurrence of new beneficial mutations is rather rare. (p.98)

Hopeful Monsters

Soon after De Vries put forth his theory that mutations were the cause of speciation and not Natural Selection, as usual a reaction set in against the idea. Richard Goldschmidt came later with a new perspective on the question of mutations. Just as Darwin focused on the Galápagos finches and De Vries on the evening primrose, Goldschmidt's work focused primarily on the gypsy moth.

Goldschmidt also disagreed with the classical theory's proposition that it was the slow accumulation of physical changes that led to a new species, and, that physical isolation is especially conducive to speciation. There is a change of phenotypes in a species as the range expands and the environment changes and these are gradual, not abrupt, and they are referred to as subspecies or races (e. g., Pennisi, 2009), but these changes may or may not be adaptive, and, do not automatically lead to speciation. The area between the old species and the new species should be an area of hiatus, but such is not the case. "There is no factual basis for the assumption that such a Mendelian polymorphism leads beyond the existence of whatever recombinations are possible" (Goldschmidt, 1940/1982; p. 138). Taxonomists always find clearly defined species. Species are separated by a "bridgeless gap" wherein there are no "missing links," *Peripatus* notwithstanding. He further challenged the adherents of the classical theory to explain how the formation of feathers, hair, blood circulation, statocysts, nerves, and other physiological structures could come about through a slow accumulation of traits. This latter is a perennial objection by many critics.

Citing experimental work, instead of gradualism, Goldschmidt put forth that speciation occurs abruptly due to changes occurring at the embryonic stage, at the chromosomal level, by external agencies and said changes affect overlapping physical traits. In some of the experimental instances, the changes in morphology exceeded the usual occurring phenotype. Therefore, he differentiated between microevolution, that is, small changes in morphology (races, subspecies) while retaining the basic species morphology, and macroevolution which leads to a new species. Natural Selection accounts for the former, but not the latter---and there is no proof that it has.

He cited the massive work of the paleontologist Schindewolf (1950/1994) who found that species occur explosively rather than regularly, as well as that of Thompson's (1942/1992) equally massive *On Growth and Form,* the latter detailing in Cartesian planes how one species can expand its form to make a new species without the need of accumulating small traits

(Goldschmidt referred both to himself and to Thompson as being ahead of their times). Scorned, he claimed that he would ultimately be vindicated (Dietrich, 2000, 2002).

Goldschmidt used the term "hopeful monsters" to refer to the mutations that he had in mind and that term has been much ridiculed by neo-Darwinists. Snickering at his name and/or his term has become almost mandatory in neo-Darwinian classes. This ridicule resulted in his theories not being seriously considered and dismissed without even looking (Gould, 1977, 1982). I, for one, admit of having fallen for this crude trick.

There's another, overlooked, element as to the hostility towards Goldschmidt, something that may sound absurd now, but at the time was very serious: he was German, attacking a theory of a leading, revered, British scientist (shades of Newton-Leibniz!), in 1940 in the middle of the Second World War. This intense anti-German hostility lasted long after the end of the war and was propagated in the movies and television in fictional characters (the villain was always German, even in a non-war plot; for example, in the book *The Satan Bug*, the villain was Italian, in the movie it was changed to German; among other things, it eliminated any suspense as to who-dun-it).[5]

FOOTNOTES TO CHAPTER FIVE

Another characteristic of successful scientists is flexibility, a willingness to abandon a theory or assumption when the evidence indicates that it is not valid.
---Erns Mayr

We think of natural selection as tuning the piano, *not as composing the melodies.*
---Jerry Fodor & Massimo Piattelli-Palmarini

[1]For instance, a logical objection was the case of mulattoes: a cross between a black and a white human results in a blending of skin color. Therefore, any new trait, no matter how adaptive, would be diluted within a population.

[2]Mendel's work is illustrative of the "You never know!" principle in science, nor was it the last. One of the fascinating aspects of basic scientific research, that is, research for its own sake, is that one never knows when those particular findings may be of use to somebody, somewhere, sooner or later. Scientific research that a particular investigator may be enthusiastic about may leave other people cold, wondering why anyone would waste his/her time on such a topic. I remember one time in graduate school, flipping through some journals and coming across just such a paper, wherein the investigator had charted the growth rate of chickens' nails and I wondered the very same thing. Yet, for all I know, that particular study may

be crucial to the investigation of the growth of cancer. In scientific research, you never know!

[3]This is as good a place as any to bring up the concept of Integrative Levels. There has been much criticism of the reductionist approach in regards to evolution, genes, behavior, etc. It is surprising that the concept of Integrative Levels, which was brought up decades ago and answers many of the criticisms, has been neglected. Simply stated, the theory states that each level of organization, be it biological, behavioral, astronomical, etc., has its own unique properties isolated from higher levels, but which must be, nonetheless, understood, and which are cumulative, so that complexity increases. See Novikoff (1945) and Feibleman (1954) for a more thorough exposition.

[4]The only light note to this were the numerous science-fiction films that proliferated in the 1950s and 1960s, which postulated the emergence of new species of animals as a result of mutations caused by atomic weapons. These invariably deadlier animals tended to be larger versions of normal animals. Humans were also mutated in these films (Baxter, 1974; Rovin, 1975; Saleh, 1979). Nothing of the sort was ever seen in Japan, or elsewhere, Godzilla notwithstanding. The same theme was found in science-fiction literature (e.g., Miller, 1959).

[5] It was not until I had finished the bulk of this book that I accidentally came into contact with both Schindewolf's and Goldschmidt's work. Hitherto, I had fallen for the neo-Darwinist trap of habitually laughing Goldschmidt away with snide remarks because of his term "hopeful monsters." *If* he was mentioned at all, it

was to snicker (which should actually have been a red flag). With Schindewolf, he was not snickered at, but neither was he mentioned---ever.

CHAPTER SIX:
BACK TO THE FIRST MECHANISM

Naturally we are all interested in facts. If when they are obtained they make the present theory untenable, the Behaviorist will give it up cheerfully.
---James Watson, *Behaviorism*

It is a good morning exercise for a research scientist to discard a pet hypothesis every day before breakfast. It keeps him young.   ---
Konrad Lorenz, *On Aggression*

Should further research discover them, we must yield to their guidance rather to that of theory; for theories must be abandoned, unless their teachings tally with the undisputable results of observation.
                    ---Aristotle.

Originally, this chapter was going to comprise a section of the following one. However, within a year or two of writing it, the concept of "intelligent design" had become prominent in the national arena (Bird, 1989; Wallis, 2005). The concept was hijacked by religious fundamentalists who, first, used it to claim that there was no such thing as evolution and, secondly, they tried to ramrod the intelligent design theory into schools' agendas with the inevitable backlash resulting; (ironically, this attempt dovetailed very nicely with neo-Darwinists' tactic of labeling anyone who questioned the classical theory of evolution as know-nothing Creationists). Ultimately, the movement towards introducing it in school curriculum was turned down by the courts; the proponents would go into court advocating it on scientific grounds while in the courtroom, then go right outside and hold a religious revival meeting. When delivering his decision, the judge noted that such hypocrisy had not gone unnoticed.

Furthermore, directly as a result of the Creationists' activities at that time, the concept of intelligent design was discredited in scientific circles, but it must be pointed out that there have been a number of theories in different fields put forth by scientists which were frankly absurd, but out of courtesy were given serious consideration (the ones dealing with the demise of the dinosaurs immediately come to mind, the most notorious being the one that proposed that the accumulated flatulence from all the dinosaurs led to their demise). The Creationists' antics effectively killed all chances for the intelligent design theory to be henceforth taken seriously and discussed scientifically. Frankly, scientists have overall paid too much attention and wasted much of their time with the Creationists. They should pay as much attention to their babblings as one would to the yapping of a Chihuahua dog.

The Principle of Irreducible Complexity

Michael Behe (1996) is a biochemist and has written a critique of the classical theory. His central argument is that, first, classical theory explains the existence of an evolutionary advanced form (either an entire organism, or its components) through the accretion of adaptive parts. But, when he examines complex anatomical structures, or physiological processes, he finds that this is not the case and the theory, therefore, breaks down.

He goes on to illustrate that the coagulation of blood is just such a process of breathtaking complexity. And if a single step is bypassed, a single component is removed, then the whole process breaks down. The process of blood coagulation is, therefore, irreducibly complex. If you skip one step, or remove a protein, the system simply collapses and there is no coagulation.

The one flaw in his entire argument, and it is a crucial one, is that he is dealing with the finished product. Coagulation in birds, for example, is quite different from humans (pet shop owners occasionally refer to birds as hemophiliacs and are very careful in periodically trimming a bird's feathers). The best approach to this problem, of course, would be to analyze the coagulation processes in the blood of humans, apes, rats, reptiles, worms, amphibians. If the evolution of coagulation can be demonstrated, it just might even make one a candidate for the nomination of a Nobel Prize. Unfortunately, it has been reported that no one does comparative physiology any more (Ehrenfeld, 1996).

Where Behe is on somewhat solid ground is in his elucidation of a cilia and a flagellum. This is because both components are a common occurrence in single cell animals and, when analyzed, do reveal that they are extremely complex structures, which would collapse if a single part is removed. So how can a flagellum, or a cilia, been gradually formed through the aggregation of individual parts---each of which, was in itself adaptive? Classical theory supposedly breaks down at this point. Darwin himself famously said as much: "If it could be demonstrated that any complex organ existed, which could not possibly have been formed by numerous, successive, slight modifications, my theory would absolutely break down." (quoted in Weiner, 1994; p. 181) As such, Behe's argument is really Darwin's---assuming he is right.

The same can be said of the various other components of complex protozoa (Patterson, 1996) such as *Stenton, Didinium* and the group of suctorials. Likewise, the very same argument can be made for any other complex physiological structure in any organism. However, one can use that argument, even though valid, only so many times before it rings hollow. To be sure, the counterargument is that there may have existed intermediary forms millions of years ago, which ultimately led to the present structure, but we lack the fossil record. As Ramón y Cajal (1887/1999) stated (p.10),

> When examined alone, the vertebrate eye or ear is a source of amazement. It seems impossible that these organs could have formed simply by the collective action of natural laws. However, when we consider all of the gradations and transitional forms that they display in the phylogenetic series, from the almost shapeless ocular outline of certain infusoria and worms to the complicated organization of the eye in lower vertebrates, not one whit of our admiration is lost and our minds are apt to accept the idea of natural formation thought he mechanisms of variation, organic correlation, natural selection, and adaptation.

Margulis, on the other hand, has a fresher approach, which is that each component of the flagellum was at one time an independent organism, or plastid, which joined together with the other components.

Behe's objection (and others') that gradual trial and error selection (Dawkins (1986) prefers the more accurate term "cumulative selection," which I agree in that it makes for a clearer picture) cannot possibly account for the interconnected complexity found in all organisms. Unfortunately, the argument between them and the neo-Darwinists has degenerated into a, "Yes, it can!" "No, it can't!" "Yes, it can!" "No, it can't!" back and forth argument, devoid of data, or experimentation, which is so reminiscent of the sterile style of arguments found in Philosophy.

Tyson (2005), on the other hand, does make a good point in his critique when he points out that, taking into consideration all the imperfections in the human body---just to take one organism---the "intelligent design" is not all that intelligent. In fact, it is downright dumb. By the same token, the criticisms of intelligent design by neo-Darwinists strikes me as a bit hypocritical. They point out, quite rightly, of many "imperfections," or, if you will, "maladaptive" traits found in many organisms and conclude that the intelligent designer—if there ever was such a creature---is not really that intelligent and thus conclude there is an "unintelligent designer." Yet, what is particularly amusing is that neo-Darwinists are constantly claiming---sometimes with an argument nothing short of bizarre in certain instances---that *all* the traits found in all organisms---by Darwinian definition---came into being through Natural Selection because they gave the organisms an adaptive advantage over its conspecifics. In short, they want to have their cake and eat it too. Even more so are what a handful of neo-Darwinists do through a linguistic legerdemain worthy of any philosopher: they *admit* that there are maladaptive traits in organisms and *embrace* those traits stating that they constitute proof of evolution through Natural Selection. How? By proclaiming that the maladaptiveness of those particular traits shows evolution in action, i.e., that evolution has not yet finished modifying those particular traits (Dawkins, 1982; Gould, 1982).

At any rate, when I originally read Behe's work, I was pleasantly surprised to read a scientific critique written in a lucid, rational, factual manner with some telling points (my first, though not my last). And then . . . at the end . . . he states that, therefore, a deity is the explanation for life and that we now have proof of "intelligent design." And he further wondered that, in light of the proof of the Irreducible Complexity principle proving intelligent design, it is strange that we are not celebrating:

> The observation of the intelligent design of life is as momentous as the observation that the earth goes around the sun or that disease is caused by bacteria or that radiation is emitted in quanta. The magnitude of the victory, gained at such great cost through sustained effort over the course of decades, would be expected to send champagne corks flying in labs around the world. This triumph of science should evoke cries of "Eureka!" from ten thousand throats, should occasion much hand-slapping and high-fiving, and perhaps even be an excuse to take a day off.
> But no bottles have been uncorked, no hands slapped. Instead, a curious, embarrassed silence surrounds the stark complexity of the cell. (p.233)

His book is very good reading. In his work, he does not throw at us quotes from the Bible as if those quotes constituted proof, which is one of those tiresome, irritating tactics of fundamentalist Christians (and Marxists!) who are simply blind to the fact that other people who are not fundamentalists (or Marxists) do not exhibit a Pavlovian salivation whenever the Bible (or Marx, or Mao, or Lenin or the Koran) is quoted. Additionally, he is upfront in saying that religion and philosophy should stay separated from science.

Nonetheless, Behe makes a crucial assumption. His argument, simplified, is thusly: there are many elements in nature which, because of the Irreducible Complexity principle, cannot be explained as having come into being through a gradual accumulation of components---therefore, a conscious, intelligent design (presumably, by a spirit) is responsible.

The *non sequitur* falls of its weight (it reminds one of Hume's retort to Descartes and the latter's testy response).[1] One does not follow from the other. Irreducible Complexity does not necessarily mean that there is intelligent design present. Simply because we may not know how a mechanism was put together, or how it first came about---whether it is a biological mechanism like a flagellum, or a true mechanism like my daughter's toy---does not mean that an unseen, immaterial spirit created it. It simply means that we do not know---yet!---how it came to be formed; it means that, like so many, many other conundrums, we have not yet found an answer.

But, we will.
We will.

The curious thing is that Behe comes close to saying the very same thing later on:

> The history of science is replete with examples of basic-but-difficult questions being put on the back burner. For example, Newton declined to explain what caused gravity, Darwin offered no explanation for the origin of vision or life, Maxwell refused to specify a medium for light waves once the ether was debunked, and cosmologists in general have ignored the question of what caused the Big Bang. (p.251)

Additionally, let me point out that---pun intended---the intelligent design hypothesis is equivalent to what is referred to in literature as a *deus ex machina*. It is as an unsatisfactory approach to literature and it is unsatisfactory to science.

He followed up his argument in a subsequent work (Behe, 2007), buttressing up his basic argument by elucidating the intricacies of hemoglobin and malaria infection (or, another way of looking at it, he repeats his argument using a different example), and, by pointing out that the ideas of common descent and natural selection are different, and, also clarifying that for him intelligent design is not Creationism and that an intelligent designer is not necessarily a supernatural being, but rather that he uses the term in a very broad sense.

It is interesting that in the brouhaha regarding the initial Behe controversy, none of the opponents really countered his principle effectively. Shapiro (1999), however, has done so.

FOOTNOTES TO CHAPTER SIX

> It is not equal time the creationists want. Don't kid yourself. They want all the time there is.
> Isaac Asimov *The Roving Mind*

[1] An identical pattern of reasoning could be found decades earlier in Erich von Dänniken, whose writings captivated the imagination of the public decades ago before scientists *finally* weighed in and disposed of his arguments. It may be remembered that his principal argument went like this: we do not know how the pyramids of the Mayas, Aztecs and Egyptians were built, so they must have been built by a technologically advanced race of space aliens.

CHAPTER SEVEN
THE SIXTH MECHANISM: SYMBIOGENESIS

If Darwinian gradualism explains the origins of animal and plant species, it follows that closely related species should have similar karyotypes. They don't.
---Lynn Margulis & Dorion Sagan, *Acquiring Genomes*

A scientific hypothesis always goes beyond (frequently, far beyond) the facts upon which it is based.
---Vladimir Vernadsky, *The Biosphere*

Honest critics of the evolutionary way of thinking who have emphasized problems with biologists' dogma and their undefinable terms are often dismissed as if they were Christian fundamentalist zealots or racial bigots.
---Lynn Margulis & Dorion Sagan, *Acquiring Genomes*

In the late 1800s and early 1900s, the classical theory put forth simultaneously by both Wallace and Darwin to account for speciation was certainly viewed by scientists as a scientific theory. That is, it was a theory that deserved consideration, should be tested and critiqued, or, amended, or, alternate theories proposed to account for evolution. At that time, the theory had not been degenerated by its supposed adherents, neo-Darwinists the likes of Ernst Mayr and Richard Dawkins, into an object of veneration, wherein criticism of the sacred object was either completely ignored or was met with a torrent of vituperation.

In fact, "Darwinism" became an object of scorn with some scientists, particularly because of "Darwinists" claiming that every single physical trait of an organism evolved because it gave that organism an adaptive advantage, something which was totally absurd on analysis. In fact, Darwin himself in *The Descent of Man*, backtracked on this point, acknowledging he had overstated the principle. Nonetheless, Darwin's adherents continued to make absurd claims---something which is still going on to this day by some individuals---claiming, for instance that the flamingos' color confused predators at dawn and dusk (Thompson, 1942/1992), or that the slanted eyes in Orientals was an adaptation to the harsh winter of Siberia---including Filipinos, Cambodians and Vietnamese.

> In the post-Darwinian era, a reaction against uncritical acceptance of the selection theory set in, which reached its climax in the great days of Comparative Anatomy, but which still affects many physiologically inclined biologists. It was a reaction against the habit of making uncritical guesses about the survival value, the function of life processes and structures. This reaction, of course healthy in itself, did not (as one might expect) result in an attempt to improve methods of studying survival value; rather it deteriorated into lack of interest in the problem---one of the most deplorable things that can happen to a science. Worse, it even developed into an attitude of intolerance: even wondering about survival value was considered unscientific. (Tinbergen, 1963; p. 417)

Therefore, we see that at the time in question some scientists were unconvinced into accepting the classical theory *in toto*, and that Natural Selection was the sole mechanism for speciation and some were even of the opinion that Natural Selection played no role at all, although they did believe in evolution. Part of the resistance was the undeniable fact that the classical theory is not amenable to experimentation and as such was viewed as useless.

- - - - - - - -
- - - - -

    Two Russian scientists, Andrei Famintsyn and Konstantin Merezhkovsky, almost simultaneously and independent of each other, conceptualized symbiosis as a major factor in evolution around the turn of the century (Khakhiva, 1992). Famintsyn noted that the Darwinian doctrine was based entirely upon indirect proofs yet thought that his theory complimented Darwin's. His primary scientific work was with lichens and cell structures, particularly the chromatophores (he preferred experimental work and disliked theorizing, a man after my own heart). Plant cells, in particular, can be subdivided into independent organisms. Famintsyn postulated that two mechanisms exist: one enables changes in an organism's morphology in order to adapt to the environment---and no more. The other is the genesis of new organisms from simpler forms.

    On the other hand, Merezhkovsky, who was proficient at theorizing, and who coined the term "symbiogenesis," thought that symbiogenesis and Darwinism were incompatible, primarily because Darwin's theory was based on old, outdated, data. His primary work was on diatom chromatophores. He was of the opinion that chromatophores were originally independent organisms that came into a symbiotic relationship with colorless cells. He postulated that both plant and animal cells arose through symbiotic combinations of independent organisms.

Boris Kozo-Polyanski was the later proponent of symbiogenesis. During the 1920s, it became evident that there was no experimental proof of Natural Selection and that the classical theory could only be supported indirectly. Acknowledging that Darwin's theory was the sole theory of evolution available at the time, Kozo-Polyanski asserted that symbiogenesis buttressed Darwinism, not undermined it; stated that symbiogenesis should not be viewed as the principal, sole, mechanism for evolution.

The universality of symbiogenesis proved that an organism was more than the sum of its parts and he collected much data to substantiate the theory, e.g., (a) the organs for digestion in ruminants, humans and insects (b) the light organs of both insects and marine organisms (c) various plants and fungi. Kozo-Polyanski asserted that Darwinism's traditional search for missing links was, therefore, a mirage.

One thing that is puzzling is that the theory of symbiogenesis was met with indifference by the official ideologues of the Marxist hierarchy. One would have logically predicted that the emphasis of mutualism, as opposed to Darwinian competition, would have resulted in an enthusiastic endorsement, yet this was not the case. This reaction, or better yet, lack of it, had both its good and bad consequences. In not being tainted by politics which would have resulted in droves of Marxist hack writers intruding into the field with their compulsive quotes of Engels, Lenin and the rest of the gang---which quotations they viewed as magical "proofs"---symbiogenesis retained its scientific integrity. One can only cringe at what might have happened if that cadaverous-looking Lysenko had turned his gaze upon this field. However, official support might have also resulted in additional bona fide scientific research. We will never know.

At any rate, technological advancements in the second half of the twentieth century would result in research that would substantiate much of symbiogenesis. But, first, a note about an American contemporary of Kozo-Polyanski named Ivan Wallin. Wallin who asserted through experiments that mitochondria and bacteria were essentially indistinguishable and that mitochondria were symbionts of animal cells; he further stated that chloroplasts also were bacteria and he introduced the word "symbionticism" to describe such a relationship. Unfortunately, Wallin worked in both a professional and geographical isolation and his work and theory were ignored and forgotten until very recently (Mehos, 1992). However, they were all preceded by Paul Portier, who claimed in 1915 that mitochondria were actually bacteria. Portier was completely ignored. He once stated, "Never will this establishment equipped with the powerful means of research that you know, forgive one isolated worker for opening a way that it should have found long ago." (Wakeford, 2001; p. 156)

The work of these Russian scientists went completely unnoticed in the west primarily due to the daunting language barrier and the intimidating Cyrillic alphabet, which was then superseded by a half century of hostile politics. But, then, so was the work of Wallin and Portier. Apparently, the *zeitgeist* was just not ready for such ideas and, as a result, science suffered.

- - - - - - - -
- - -

Lynn Margulis, like Barbara McClintock, was one of those women scientists who were told that their findings could not possibly be true only to ultimately be proven right and everyone else wrong. But, unlike McClintock she had a facility for proselytizing outside of science through readable, nontechnical, language and pithy zingers (Margulis, 2004). Her seminal ("painfully convoluted, and poorly written" (Margulis, 1998; p. 29) paper on her theory of endosymbiosis was rejected by 15 journals before finally being published in 1967 in the *Journal of Theoretical Biology* through the personal intervention of the journal's editor (Margulis, 1998; Schaechter, 2012). She put forth the proposition that the mitochondria, plastids and flagella were independent organisms that had been incorporated into cells and existed through symbiosis, rather than what been previously assumed, namely, that mitochondria were simply structures that were part of the cells, such as the Golgi complex, the endoplasmic reticulum, or the centrioles (Sagan 1967). Although it has been known very early on that lichens and sponges were symbiotic organisms (Vogel, 2008), thanks to great technical advancements in science, it is a generally accepted fact that mitochondria are, indeed, similar to bacteria, not only in appearance, but in biological processes (Raven, 1970). Ultimately, she was conclusively vindicated when the evidence started pouring in, including from previous, overlooked, research. Nonetheless, she did experience resistance; one of her grant applications was rejected with the message, "Your research is crap, do not bother to apply again" (Sagan, 2012), possibly by a neo-Darwinist.

In *Acquiring Genomes, A Theory of the Origins of Species,* Lynn Margulis and Dorion Sagan (2002) provided a more comprehensive theory of serial symbiogenesis, thanks to up to the minute research findings in many areas of biology (what was particularly bizarre in their book was that it sported a foreword by Ernst Mayr). They pointed to the fact that most scientists admit that mutations (the cornerstone of species variability and, therefore, evolution) are 99.9% deleterious and have been overemphasized by neo-Darwinists.

From coral to insects to clover to humans to ruminants to squids to sponges to lichens to Portuguese man of wars to plant cells and animal cells, symbiosis between different species is an indisputable fact (O'Neill, Hoffmann & Werren, 1997; Gilbert, et al., 2012). One protozoa is actually composed of five organisms (Margulis, 1971, 2008). It appears that the majority of legumes and insects are dependent on bacteria for their nourishment (Wakeford, 2001). But it is not just simple organisms. Vertebrates, including humans, are composed of intracellular symbionts, and system-wide symbionts (cows and humans would probably starve without the intestinal flora). As such, "serial symbiogenesis" is the originator of species (it is curious that although a great critic of the classical theory of evolution, Margulis claimed that her theory did not supplant Darwin's but simply appended to it,[1] pointing to, among other things, the undisputed fact that no new species have occurred, with or without human-induced mutations: "Many new ways to induce mutations are known but none lead to new organisms" (Margulis & Sagan, 2003; p. 11). Serial symbiogenesis has the additional advantage that it can also point to experiments that support it (e.g., Nass, 1969).

The fact that bacteria are so notoriously promiscuous in the exchange of genomes back and forth that some scientists have put forth the proposition that there is no such thing as a species when referring to bacteria (Wakeford, 2001). Indeed, it is fascinating to learn that 250 of the 30,000+ human genes come directly from bacteria. Another report (Gladyshev, Meselson and Arkhipova, 2008) found that rotifers have genes which originated not only with bacteria, but also fungi and plants. Microbiomes are partially inheritable (Mizrahi & Kokou, 2019) and Corbicular clams steal genes from other species.

However, what I feel is even more fascinating is that the microbiome has been found to affect the behavior of some animals (Lyte, 2013; Foster & Neufeld, 2013; Ezenwa, et. al, 2012).

Incidentally, Margulis was to ultimately discover the work of the Russian scientists and became instrumental in introducing them to the West.[2]. However, it appears that she initially overlooked Reinheimer

(1915/2012), Barricelli (Hackett, 2019) and J. E. Wallin (a physician, not a biologist) who predated the theory of symbiogenesis as the basis for evolution. Another theory of evolution has recently been based on symbiogenesis (Zilber-Rosenberg & Rosenberg, 2008).

At any rate, the reader has undoubtedly already made the leap to Lovelock's Gaia hypothesis (1988), a point that Margulis and Sagan do not fail to make. Like Margulis, Lovelock received a lot of resistance and scorn for his theory. And, like Margulis, Lovelock would ultimately learn that his theory had a Russian predecessor, in the work of Vladimir Vernadski (1926/1997).

The theory of symbiogenesis as put forth by Europeans a century ago and restated and updated by the Americans is an exciting one. It is also a convincing one which, like all theories, await further testing and criticism, although it must be pointed out that much data already supports it. That its proponents welcome such testing and critiques is a noticeable advantage over Darwinism as viewed by the Torquemada of neo-Darwinism, Richard Dawkins.

However, as Kozo-Polyanski himself asserted, symbiogenesis cannot be the sole mechanism for speciation. It seems to be applicable up to a point in the evolutionary scale (although there is one case of plastid intracellular invasion of a vertebrate (Kerney, et al., 2011) as opposed to the instances with invertebrates (Geddes, 1882; Pierce, Biron & Rumpho, 1996; Maeda, et al., 2012; Somvanshi, et al., 2012) and plants (Richardson & palmer, 2006; Bock, 2009)). Without the microscopic symbionts, cows, sheep, millions of other species would probably cease to exist, but what Margulis and others have failed to explain is how, exactly, do new species arise, particularly in the case of vertebrates. Something else, then, must come into play with higher organisms.

Nonetheless, I will end with yet another quotation from Margulis and Sagan (2008; p. 82):

> The heavy hand of selection can and does dramatically change the proportion of heavy egg-laying hens, grape-sized versus cherry-sized tomatoes, or long-beaked ground finches. But no one has ever shown that this process does more than change gene frequencies in populations. Intraspecific variation never seems to lead, by itself, to new species.

FOOTNOTES TO CHAPTER SEVEN

[1] This insistence of buttressing the classical theory instead of supplanting it was echoed by Gould and Eldredge when they put forth their punctuated evolution view.

[2] It is rather fitting that Margulis (1998) also championed an end to "academic apartheid" (the fact that scientific disciplines do not interact with each other and are ignorant of each other's work). This would be a sort of academic symbiogenesis.

CHAPTER EIGHT
THE FLAWS IN THE CLASSICAL THEORY

A cat is not a dog.
---T. S. Eliot, *Old Possum's Book of Practical Cats*

A beautiful theory, killed by a nasty, ugly little fact.
---T. H. Huxley

Something isn't right in the theory of natural selection.
---Tijs Goldschmidt, *Darwin's Dreampond*

Bacteria may be morphologically simple, but biochemically they are exceedingly complex.
---Simon Conway Morris, *The Crucible of Creation*

The classical Darwinian-Wallace theory of evolution, over a century old and essentially intact, is deeply flawed. Although some biologists occasionally exult at some recent phenomena or other that they feel substantiates the theory, others are not so sure (Coalacino, 1977). A few biologists and paleontologists (and a few well informed members of the public, for that matter) feel that there is something . . . not . . . quite . . . right. They are not sure just what it is, but they know that something is wrong. But they only whisper it. They have been intimidated. The only ones that are blissfully happy with the status quo are the neo-Darwinists, who, like the alcoholic, are in a state of blissful Denial. But, like the person who is an alcoholic, the first, and hardest, step is to admit that there is a problem.

> As students are quick to point out to their teachers, the argument of natural selections is very nearly circular. In its circular form it says: (1) The more fit genotypes leave more descendants which, because of heredity, resemble their ancestors. (2) "Fit" genotypes are those which leave more descendants. (MacArthur & Wilson, 1967; p. 145)

The question of speciation remains a mystery. If we use an analogy of a jigsaw puzzle, then what we have here is a puzzle whose subject matter, say a beachscape, has become evident. We have numerous pieces, particularly on the periphery, so that we can tell that the jigsaw puzzle is that of a beachscape. We see the waves, the sky and clouds, the seagulls, the dunes, some crabs and sea oats. But there is a central piece missing, a crucial piece in the center, much larger than the other pieces, that will finally make overall sense of the whole Gestalt and explain away the apparent contradictions. To use our analogy further and use a little symbolism, the center piece is that of a lighthouse. What, in reality, is that missing piece? Where is the lighthouse? At this point in time we cannot see it, we are not certain of it, though a century from now it will be common knowledge (and to those persons then it will seem self-evident, and puzzling to them that we were not conscious of it). Only the neo-Darwinists, with their tunnel vision, insist that the jigsaw puzzle *is* finished, *is* complete, with no pieces missing, and they do so happily.

Let us now examine some of the *data*.

The Arms Race

One of the annoying clichés in evolutionary theory (and how the field is filled with clichés!) is that of a constant arms race between prey and predator. According to the cliché, whether it is plants or animals, prey are constantly evolving defenses to fend off predators, either in developing longer and sharper thorns in plants, or in speedily running away from a carnivore (Diamond, 1987; Weiner, 1994; Stahl, Dwyer, Mauricio, Kreitman & Bergelson, 1999). But, is this truly accurate?

There are several flaws in this kind of reasoning. First, almost every physical, or behavioral, trait that can be quantified, in terms of size, speed, age, response time, etc., invariably falls into a bell shaped curve. This is true for all prey and all predators. In the case of wildebeest, or the Thompson's gazelle, each animal's top speed, compared to others' in the herd, and to the individual's day to day performance, falls somewhere along the normal distribution scale (as is the case with racing horses, or human track athletes, or, the length of leaves in plants or plants' overall height). Likewise, a lioness, or a cheetah's, speed and duration of pursuit, when plotted against others in the pride, or other prides, or, for that matter, with that very same individual at that particular month, also falls somewhere within the bell shaped curve. The curve may demonstrate kurtosis, to be sure, but it is still a curve. So that, barring incidentals, such as confusion, illness, or surprise, a cheetah will seize the slower gazelles. The slower gazelles and impalas get eaten, their progeny die out, and everyone moves up the scale, so that the average speed of gazelle is constantly increasing. As the gazelles and impalas get faster, slower cheetahs die out from starvation until only the fast cheetahs remain. And the process continues forever after.

Yet, that is not the case. If the arms race was, indeed, taking place, both cheetahs and gazelles would be, by now, a near invisible blur. Their respective speeds would not be plotted as bell shaped, but columnar and we all know that this is not the case. Further, the bell shape remains constant and it remains constant around a constant mean, not only for the herd/pride, but for the individual itself.

Peppered Moth

Another annoying cliché is that of the Peppered Moth, so well-known that it has even wormed its way into an astronomy book, of all places (Jastrow, 1967).

As the litany is usually presented (Huxley, 1953; Weiner, 1994; Eldredge, 2001; Mayr, 2001; Parker, 2003), the moth (*Biston betularia typica*) used to rest on the lichen-covered trees of Great Britain. Because of its coloration, it blended right in and was overlooked by avian predators, due to its natural camouflage. But, because of enormous pollution in the British midlands, the bark of trees became solid black, so that the moth stood out against the dark background. Simultaneously, a dark variety of the moth, *B. b. carbonaria,* initially rare in occurrence, now blended right in and multiplied in numbers. The story is usually accompanied by pictures of birds eating those moths. The incident is then triumphantly proclaimed to be another example of evolution.

The problem is that it is wrong. In every respect (Coyne, 1998; Kruuk, 2003). First, the moths do not rest on trees (where they do happen to rest at all is still a mystery). Second, in the original "experiment," the moths were released in the daytime; as the reader is undoubtedly aware from personal experience, moths are practically catatonic in the daytime. Third, the moth does not have a tendency to choose a matching background in which to rest. Fourth, and more importantly, in the United States, a similar change in distribution of both subspecies of moths occurred---even though no comparable change in the environment to that of Great Britain took place; in other words, it was a control group. The assumptions regarding the Peppered Moth is similar to the assumption made of marine fish with an "eyespot" on their dorsal area which was assumed to divert predators, but which turns out not to be so (Gagliano, 2008).

But, still, let us assume for the sake of argument that the case of the Peppered Moth is a valid example of evolution, as it is usually proclaimed. What---exactly---does it prove? "Evolution"? But what do they mean by evolution?

One comes across many such assertions in the literature as, for example, when red tide wipes out an existing marine ecology. When new species move in to occupy the vacated niches, it is, likewise, gleefully proclaimed as an example of "evolution at work."

In the case of the *Biston* moth, as any Behaviorist would have pointed out, the only thing that happened was that one existing variety of moth became more numerous than another. A new *species* was not generated by the change in the environment.

So, before we proceed any further, let us clarify

one thing which should have perhaps been clarified much earlier. The real question of evolution as far as this book is concerned, is not about the origin of life, or adaptation, or survival of the fittest, or of the domination of ecological niches. The topic of this book, and for that matter the topic of evolution itself, is, quite simply, the topic of *speciation*. How do new species arise? That is the core topic of evolution, the sole area of investigation. This narrowing down will eliminate confusion as to whether, or not, evolution is occurring in the world.

The Finished Product

All species are distinguished from other species by a number of physical and, in the case of animals, behavioral traits. By definition, as put forth in the classical theory, all the diverse physical and behavioral elements that make up and define an organism---each and every one of those elements!---came into being through Natural Selection. They were selected for because they conferred on that organism a distinct advantage over its predecessors. Those elements---in each and every organism---differentiate that organism from other species, and, from its predecessor.

Yet, it is totally absurd to think that every color, every pattern, every shape of crest, or feathers, or paws, or skin, or eyes, or hair, has come into existence because it bestows that particular species with an advantage towards survival. That is just plain absurd.

They are incidentals.

But if each and every one of those elements did not arise through Natural Selection, then just how did they arise? And why?

OWLS

Taking species as a whole, no one can say just why the appearances, i.e., the physical differences between a Short-eared Owl (*Asio flammeus*), a Long-eared Owl (*Asio otus*), a Great Gray Owl (*Stryx nebulosa*), a Spotted Owl (*Stryx occidentalis*), a Barred Owl (*Stryx varia*), a Northern Hawk Owl (*Sturnia ulula*), a Great Horned Owl (*Bubo virginianus*), a Flammulated Owl (*Otus flammeolus*), a Common Barn Owl (*Tyto alba*), why their respective appearances, either in overall appearance, or detail by detail, confer an adaptive advantage over each other.

To be sure, the Snowy Owl's (*Nyctea scandiaca*) plumage confers the advantage of camouflage in the northern regions. And the Florida Burrowing Owl (Speotyto cunicularia)'s fixed action pattern of digging and nesting underground can arguably be called advantageous, especially in some of the treeless islands off the peninsula; likewise, its diminutive size allows it to survive on smaller, more numerous prey such, as insects and lizards, rather than compete with the relatively less numerous, though larger, prey.

But no one can claim---even by a stretch of the imagination---that the two tufts of hair above the head of the Great Horned Owl bestows a survival advantage over, say, the Barred Owl. They are incidental. And that is one of the Achilles' Heels of the classical theory as it now stands, for the theory implicitly states that what differentiates one species from another---indeed, why another species arose from a previous one in the first place---is an accumulation of traits, each and every one of which was advantageous towards the survival and propagation of that species. If there is no advantage whatsoever to a trait arising in an individual, then that particular organism will not become more successful at surviving and, therefore, become more numerous and, ultimately, dominant.

However, if an organism does acquire a superior trait, and, it has numerous progeny, and, for the sake of argument, becomes a distinct species, then why does a secondary, incidental, trait, which is not superior in itself in any way, become uniform throughout the new species?

After all, of what adaptive advantage does the tuft of hair at the end of a lion's tail entail?

Rather, it appears to be more logical that a species comes into being as a package deal, with a combination of unique traits that differentiate it from the other species, some of those traits possibly, though not always, conferring an advantage over other species and with other traits having no advantage whatsoever; in the latter case, the organism could take them, or leave them, as it were, with no noticeable change in its existence. Furthermore, there is great variation within the species as to those existing traits (including size, number of stripes or spots, shade of color, etc.); variations within traits do not simply vanish, as any naturalist can testify.

The basic problem is that biologists at times anthropomorphize Nature when they ask, "What is this trait good for?" A trait may not have any particular utilitarian function, although some traits may have an occasional, mild, non-crucial advantage. It just *is*. It has no ulterior purpose that we have to really, really try hard to figure out. What truly matters are *the internal biochemical processes*, and that tends to be pretty consistent from organism to organism, anyway.

> In other words, features that taxonomists typically end up using to separate species, especially very similar species, from one another are not vital, as best as can be determined, to the existence of the species. Rather, the characteristics taxonomists often use to distinguish even very similar species from one another are seemingly useless and trivial. Echoing the general sentiments of de Vries, Bateson had to conclude that species differences are not those of adaptation. (Schwartz, 1999; p. 203)

Of relevance here is the advantage of sexual, as opposed to asexual, reproduction. Biologists are at loggerheads as to what, exactly, makes sexuality so much more advantageous to organisms than asexuality (Tindol, 2001). No. It just *is*. Why has asexuality survived and proliferated if it is so disadvantageous? Rather, sexuality was a trait that was developed early on and continued to be passed on with further speciation, just as a rib cage has been passed on to all organisms having a skeleton, even though it leaves the rest of the abdomen exposed to injury. Margulis & Sagan (1997, p.86): "Actually, sexuality is not required for reproduction in most members of four out of the five kingdoms of living things." And more to the point (p.163):

> Biologists have thought that sex persists because it increases the variety, the newness of offspring. This variation, it was reasoned, allows sexual organisms to adapt faster to changing environments than do asexually reproducing organisms. Yet there is absolutely no evidence that this is true.

Carnivorous Plants. Let us now consider the insectivorous plants, beginning with the pitcher plants.

In eastern North America, they belong to the *Sarracenia* genus. There is a great variation in the morphology of the leaf-trap within the genus. The exact number of species fluctuates by one or two, depending on which botanist you talk to. With the exception of one, *Sarracenia psittacina,* the Parrot Pitcher Plant, the leaves grow vertically. For our purposes, we shall consider only three, *Sarracenia flava,* the Huntsman's Horn, *Sarracenia rubra,* the Miniature Huntsman's Horn and *Sarracenia purpurea,* the Purple Pitcher Plant. A pitcher plant eats insects, albeit in a passive manner, unlike other insectivorous plants, such as the sundews, the bladderworts, the butterworts, the Venus' Flytrap. Emerging from rhizomes, the leaves develop into hollow tubes, usually colorful to some degree. Near the entrance, glands produce drops of nectar which some insects are seen to eat. Presumably, the nectar exudes an attractive odor to insects and presumably the colorful leaves may give the appearance of large flowers. Once at the entrance, insects tend to wander inwards, since other drops of nectar can be detected. However, inside, the upper lining of the leaf is extremely smooth. Not only does it look smooth to the naked eyes, but insects trying to enter can be seen tentatively trying footholds, which was not the case on the outside. Inevitably, they slip and fall in. If they don't slip, the insects become drunk from the fermented nectar and they fall in. The inside of the leaf, however, is also lined with downward pointing hairs (they have been compared to antitank obstacles) which further prevent the insects from climbing back out. Near this region glands secrete an enzyme which digests the insects, the nutrients are absorbed and the accumulated, lifeless, exoskeletons pile up until the leaf eventually dies off and is replaced by another leaf (oftentimes, a spider can be seen at the opening of the leaf in wait of an insect, apparently having learned that in that spot, it is easy pickings, or, more likely, it can detect the nectar).

In regard to evolutionary theory, the question now becomes: what possible advantage does one species have over the others that it evolved away from the ur-*Sarracenia*, and, for that matter, which one came first, *S. rubra, S. flava,* or *S. purpurea*? Which came first, the chicken, or the egg? goes the old joke, but it is no joking matter here. Their habitats often overlap. As a matter of fact, naturally occurring pitcher plant hybrids can be occasionally encountered in the wild (rarely in animals (Gompert, et.al., 2006)). One has no real advantage over the other. True, the Huntsman's Horn is much bigger (3') than its miniature counterpart, whose diet is mostly ants and gnats, but . . . so what? Both thrive well. And, besides, did the Huntsman's Horn evolve from the Miniature Huntsman's Horn, or vice versa? Or did both evolve from one of the other species, say *Sarracenia minor*? Or, perhaps from some extinct ur-pitcher plant? True, the *Sarracenia purpurea* does have one advantage in that its range of territory extends from the southern Atlantic seaboard all the way to Canada, far beyond the warmer climate, where its conspecifics thrive, but, why the change in the leaf's morphology, if it is just a matter of thriving in a wider range of environment?

Then, there is the question of the plant's

evolutionary morphology. If one accepts orthodox theory, a species comes into being through the slow accretion of unique traits. So how did a pitcher plant come into being? What, exactly, was the sequence of events that ultimately rendered the end product a pitcher plant? First, tubular leaf, then nectar glands, then hairs, then hollowing of leaf, then coloration, then opening of leaf, then smoothness, then glands with enzymes, then creation of enzymes, then excretion of enzymes upon stimulation, then absorption of nutrients, then distribution of nutrients? Or . . . first gland with enzymes, creation of enzymes, then excretion of enzymes upon stimulation, then tubular leaf, then coloration, then nectar glands, then smoothness, then absorption of nutrients, then hairs, then hollowing of leaf, then opening of leaf, then distribution of nutrients. Or . . . .

Yet, one can sense that this is intuitively wrong. The entire leaf is "designed"---perhaps "structured" is a better word, "designed" is too anthropomorphic---for two functions: photosynthesis and entrapment. Is it not more likely that, instead of a sequence of nonadaptive traits succeeding one another---each of no benefit whatsoever to the plant---rather, the plant came into being as a whole with most of its basic traits already in place?

Additionally, note that by breaking up the components of a pitcher plant being created step by step, although it may seem complicated having done so, it is actually an oversimplification. The biochemical processes for activating and carrying out each step (e.g., the creation of enzymes, the secretion of enzymes, the detection of just the right stimulation (a bug, not a pebble) are terribly complex and I will not go into them partly because it would take us away on another long alley and partly because some of the processes, in my view, have not been as thoroughly analyzed by botanists and biochemists as they merit since Lloyd (1942).

Let us now also examine another carnivorous plant, the Venus' Flytrap, *Dionaea muscipula* Ellis, which is better known. The leaves emerge in a rosette form from a bulb. At the end of the leaf the trap slowly develops. Then, it opens. Things happen. First, the inside of the trap reacts to the sunlight and turns the inside into a pink-to-deep red color. Second, at the rim of the trap, nectar glands produce droplets which some insects evidently eat. There are three stiff black hairs on either side and if any two are bent within a certain, brief, interval, the trap closes, sometimes slowly, sometimes rapidly.[1] If an insect, or a piece of meat is dropped in, triggering the hairs, the sides of the trap close and begin to squeeze it; if it is a pebble, there is no squeezing and the trap reopens the next day. Otherwise, the insect is digested, the trap reopens and the leftover chitin blows away. After three successful entrapments, the leaf dies and is slowly replaced by another (if a piece of meat is dropped without triggering the hairs, the sides of the trap begin secreting enzymes and begin to bend towards the stimulus, a carnotropism)[2].

Incidentally, before the trap begins to squeeze for a few seconds there are tiny gaps between the spines at the edge of the trap and one person has pointed to this particular, transient, feature as "an evolutionary adaptation" that allows tiny, insignificant insects to escape. Utter nonsense! First, such a small insect could not bend the trigger hairs to activate the trap and, second, the real danger to the trap comes from capturing prey that is too big. In the latter situation, it inevitably results in the leaf being overwhelmed by rot.

There is only one genus and only one species and I choose it in particular because of its morphological and geographical isolation. We must now ask again, how did the Venus' Flytrap evolve into this plant? The Venus' Flytrap, according to orthodox theory, came into being through a steady progression of accumulated characters which gave it an advantage over others of the original species through Natural Selection. But this argument breaks down if we do just that, if we break the organism down into its unique predatory components. Consider the discreet components (in no particular order):

- nectar glands
- exuding nectar
- chemical composition of nectar
- growth of trigger hairs
- electrical triggering by bending of hairs
- shape of hairs (they are both flexible and indented where they bend)
- spines
- trap closure
- carnotropism
- enzyme producing glands
- morphology of trap
- initial opening of trap
- chemical composition of enzymes
- absorption of nutrients by trap
- distribution of nutrients

plant   -absorption of nutrients by the overall

And, again, keep in mind that behind each of the overall characteristics lie *complex biochemical processes* though we will not descend into the microscopic level and examine the incremental processes. It is one thing to simplistically say, "The traps developed in order to obtain additional nutrients," it is another to, step by step, assemble the biochemical details of such a mechanism.

We can now be a little absurd, just as we were with the *Sarracenia* pitcher plants, and truly randomize the sequence of development of the trait, as such:

        -first, there was carnotropism
        -second, came the spines
        -then absorption of nutrients
        -then coloration
        -then nectar glands
        -then chemical composition of nectar
        -then enzyme producing glands
        -then chemical composition of enzymes
        -then morphology of trap
        -then growth of trigger hair
        -then indentation of trigger hair
        -then electrical triggering by bending hair
        -then exuding nectar
        -then exuding enzymes
        -etc.

But, as we said, this is absurd randomization. Natural Selection works through a cumulative process, each step being advantageous to the organism. Fair enough. Where do we start? Nectar glands. We can start with them. Nectar glands will attract pollinating insects which will benefit the ur-Venus' Flytrap (though not really to the extent that other plants will be at a disadvantage, not really). Very well. Now, feeling generously, though going against classical theory, we (somehow) add at the same time both the glands, the chemical composition of nectar and the exuding of nectar on the leaf, all of it as a package deal and we pretend that that gives the plant a competitive edge to it (somebody please say how). What then? Well, now we say that the plant evolves the shape of the "trap" and we pretend---again---that that gives the plant a competitive edge to it (somebody again please say how). Next, graciously, we add coloration to the trap. Together with the nectar glands---which we very conveniently situate at the inside border of the prototrap---it helps to attract pollinating insects, just as if it were a flower (so that pollinating insects are drawn away from the real Flytrap's flower?). All right. Now what? We add hairs. Inside the trap, not outside. And indented. Why have hairs been selected for? We do not know. We really do not know. Then, becoming impatient, in one fell swoop, we add electrical signaling upon the hairs being stimulated *and* closure of the trap. Now, we will truly have a curiosity akin to *Mimosa pudica*, the Sensitive Plant. And what advantage does the small plant obtain by having this mobile curiosity? For the life of me I cannot think of one.

Not one.[3]

All right. We now have nectar, coloration and trap closure. All we have to do is "simply" add spines at the edge of the trap. That one additional trait is definitely a product of Natural Selection, right? No. And last, but definitely not least, we add enzyme glands (which we pretend is a unitary trait rather than a complex, multifaceted structure). We also, feeling generous, in a comprehensive gift, we add the combos of enzymes and the secretion of enzymes.[4]

Is there anything else that we left out? Yes! We forgot to absorb the nutrients.

And distribute the nutrients.

And have the cells incorporate the nutrients.

And open up the trap once more.

And voila! We now have the finished product, *Dionaea muscipula* Ellis, the Venus' Flytrap, courtesy of Natural Selection.

What could be simpler?

Now---to use the modern parlance---what's wrong with this picture?

Barthlott, et al., (2007) attempted to hypothesize how carnivorous plants came to be, but were unsuccessful.

> When examined alone, the vertebrate eye or ear is a source of amazement. It seems impossible that these organs could have formed simply by the collective action of natural laws. However, when we consider all of the gradations and transitional forms that they display in the phylogenetic series, from the almost shapeless ocular outline of certain infusoria and worms to the complicated organization of the eye in lower vertebrates, not one whit of our admiration is lost and our minds are apt to accept the idea of natural formation thought he mechanisms of variation, organic correlation, natural selection, and adaptation. (Ramón y Cajal, 1887/1999; p.10)

This is not the case with carnivorous plants.

Obviously, the example that I used with these plants (and plants are rarely alluded to when discussing evolution, it is almost always animals---very curious) is like the argument that both Darwin (the eye) and Behe (the flagellum), writ large. Holistically, rather than reductionist.

One other point: in viewing the inextricable interconnection of elements in complex phenotypes, it is understandable how seductive is the illusion of intelligent design.

Living Fossils. To recap: one facet of evolutionary theory states that evolution is an ongoing, active process, one that becomes evident only after--- decades? Centuries? Millennia? Eons? It is never truly specified.

All individuals differ from each other to some degree. From time to time, certain traits in an individual organism arise which happens to give that particular organism an advantage---whether it is in obtaining nourishment, avoiding predation, obtaining a mate, or persevering through harsh environmental conditions. This advantage will supposedly result in that particular organism ultimately becoming more numerous as well as becoming a new species, while the previous form of that organism dies out because it suddenly becomes maladaptive. Presumably through inbreeding, the descendants of that one lucky individual will supplant the original stock.[5] Human beings approximate this process in a shorter period through selective breeding of certain plants and animals (livestock and pets).

In this respect, the classical theory is wrong. With this rationale, the *only* way that a species would not evolve is if all its members were genetically identical, i.e., if they were clones. And yet, we see many species that, after *millions* of years, have remained *unchanged.*

Coelacanths. I first came to know of coelacanths in my early teenage years through a delightful little book, *The Maybe Monsters* (Soule, 1963), which, although out of print, I heartily recommend for youngsters, if you should ever come across it in a used bookshop.

The story of the discovery of the living specimens of coelacanth is a fascinating one, unusual in that women played such a crucial role throughout (Thomson, 1991; Fricke & Hissmann, 1990; Erdman, Caldwel & Moosa, 1998, 1998; Forey, 1998). Very briefly: Coelacanths were known only from fossils as having lived around the time of the dinosaurs. In December 1938, Marjorie Courtenay-Latimer, a young curator at a small museum in South Africa, having established a network of contacts among the fishermen, obtained a strange large fish, which she immediately recognized as being unusual. While trying desperately to preserve it during the summer heat with practically no means to do so, she wrote to an ichthyologist friend of hers 350 miles away, enclosing a crude drawing. L. B. Smith was so puzzled by the drawing that he just stared at the letter until his wife[6] brought him to. Smith spent the following month trying to obtain information on fossils and approaching his colleagues on the subject, who rebuffed him. Meanwhile, the oily fish was rotting away and Latimer had no choice but to have it mounted. The taxidermist, unfortunately, threw away the viscera, and to make matters worse, the photographs that she took were ruined by the developer. She sent Smith some scales which finally confirmed his suspicion that it was a member of the extinct Order of the Coelecanthini that had lived 80 million years ago and he and his wife finally drove the long distance to see the now mounted specimen for themselves. He officially named it after her, *Latimeria chalumnae*.[7] The event was headline news around the world. In the hopes of obtaining another specimen, Smith distributed fliers announcing a hefty reward for another specimen. While he waited, he wrote a number of papers on the subject for the scientific journals. Then, in December 1952, after World War II, he received a telegram from the Comores Islands informing him of natives having caught another fish. He urgently cabled the South African Prime Minister requesting a plane; by coincidence, the Prime Minister's wife happened to be reading at the time one of Smith's books and a plane was put at his disposal and Smith was able to bring it back. Again, a third specimen was caught in September 1953 in the Comores but, this time, the French authorities banned all foreign scientists, claiming that the coelacanth was a French coelacanth, having been caught in the French Indian Ocean off the French Comores Islands, (presumably eating French food). Smith was not allowed to go near it, or any of the subsequent specimens, as French scientists proceeded with their own private examination.[8]

Switch suddenly now to 1998. One Mark Erdmann and his bride, Arnaz, were honeymooning in the island of Sulawesi (Celebes), in Indonesia. As they strolled through the local fish market, Arnaz spotted, of all things, a coelacanth being carried away and pointed it out to her husband, but it was not until later that Mark truly realized what they had seen. He got a second chance after he contacted the National Geographic Society for emergency funds and was able to obtain another specimen---caught by the same fisherman. While the specimens from Comoros have been blue-gray, the Indonesian ones are brown. Otherwise, they are practically identical. Note: the coelacanth are in two different regions, a third of the world apart.

Horseshoe Crabs. In my cabinet I have a fossil from the famous Solnhofen limestone in Germany from the Jurassic era, 190 *million* years ago. It is unmistakably the fossil of a horseshoe crab, the same animal that I have occasionally encountered in Florida, either when they spawn, or swimming in shallow waters by stroking their legs in unison (I jokingly call it the last surviving trilobite, which is, of course, wrong). There is no mistaking it; the fossil and the contemporary one are one and the same, unchanged through millions of years. It is an exceedingly primitive and odd-looking organism, its compound eyes in particularly having been thoroughly studied.

Echinoderms. Also in my cabinet is a black fossilized sea urchin from the Pennsylvanian era, 294 *million* years ago. It is an unmistakable sea urchin, caught as it died, with its spines declining and detached (a common occurrence in my saltwater aquariums). For all I or anyone else knows, the particular fossilized species may still be alive.

Likewise, I have a brittle star, again from the Solnhofen, fossilized during the Jurassic, 190 *million* years ago. If I make a sketch of it and of a contemporary, living brittle star, I defy anyone to identify which is the fossil and which is the living one.

Additionally, I have several samples of sand dollars, some in matrix, others free standing, which are indistinguishable from modern sand dollars. So indistinguishable, in fact, that one purveyor of fossils when showing his inventory to me got mixed up as to whether a particular sand dollar was, indeed, a fossil or one of his dried, modern specimens. The sand dollars are not as old, having been fossilized "a mere" 20 to 50 *million* years ago.

Other Living Fossils. I hold in my hand a fossilized shrimp from the Cretaceous era, dug up in Lebanon, probably the same kind of shrimp that I had last night for dinner. In my other hand I hold a chunk of Colombian amber with numerous insects trapped inside, similar to living species (just how many insects are, indeed, living fossils can be seen in Grimaldi's and Engel's (2005) book). And in front of me, there is another fossil from Lebanon from the Cretaceous era of either a crawfish, or a lobster, probably the former.

Something that I do not have are fossilized starfish, nor a living purple frog ("the coelacanth of frogs" as it has been called) which was recently discovered, a living fossil and which had been thought extinct as of fifty *million* years ago (Aggarwal, 2004).[9] Also recently found, off Brazil, is a Chimaera fish, which was around 150 *million* years ago (Astor, 2004). Nor do I have the algae known as *Bangia*, which has been around since *1,200 million years ago*, way, way before the dinosaurs (Butterfield, 2003). Nor, for that matter, do I have the *Voltinia* butterfly, still found in

Latin America, which is 20 *million* years old (Vane-Wright, 2004). Nor do I have the slug-like *Neopilina.*

They may not be as dramatic as a coelacanth, but they are dramatic because they are still around. I also do not possess a Wollemi pine tree, which is understandable because there are only forty living specimens and the particular microclimate in which they were found in, in an Australia gorge, has been wisely kept secret (Parker, 2003). And I would like to at least see lungfish, which are as old as the Permian (possibly even the Carboniferous and Devonian) since two of the seven families of fossil lungfish survived into the Triassic. And, recently, a frilled shark was filmed swimming off Japan, and another caught off Portugal, a living fossil. And, the New Zealand archipelago boasts of the tatuara lizard, the large Weta insect and the microscopic Protulophila.

But by no means are those all of the living fossils! Ferns, horsetails, clubmosses, corals, arowanas, *Stygobromus canadensis*, hagfish, lampreys (from the Carboniferous era), aardvarks, *Neopilina galatheaem* sturgeons, lancetfish, *Cymatioa cooki*, ghost sharks, alligator gar, bisons, crocodiles, turtles, arapaimas, *Polystoechotes punctata*, *Monoplacophoras, Welwitschias*, tadpole shrimps, *Monoplacophorans*, Cuvier's bichirs, *Lingula* (since the Cambrian!), water bears (also from the Cambrian---and 1500 species of them!) are all living fossils (the jury is still out on a specimen that may be a Cambrian graptolite). (When can we expect them to evolve?) One could even make the case that hippopotami, tapirs, camels, pigs, horses and rhinoceroses are living fossils.

And, to indulge in a bit of harmless, yet wild, speculation, with absolutely no hard data whatsoever to back it up, in regard to the ancient cultural icons of dragons in Europe and Chinese mythology, could they be due to a true living fossil, a dinosaur, once encountered by early humans in some ecological niche, somewhere in Eurasia, and subsequently exterminated? Perhaps the Megalania (*Varanus priscus*)? Carl Sagan (1977) also wondered about the ubiquity of the dragon legend.) Or perhaps this was simply due to human beings encountering fossil bones and letting their imaginations run wild (Mayor, 2005).

And then, there are the stromatolites; for 90% of the Earth's life, they ruled the Earth---90%! *Ever since the Precambrian!* Even before trilobites!

Unchanged.

But for that matter, we can just as well point to unicellular organisms like bacteria. They are definitely living fossils. Andrew Knoll (2003) makes this very same observation (p.110):

> Cyanobacterial fossils seen earlier in cherts from Spitsbergen and the Belcher Islands hint at a general, and remarkable, feature of this group---populations preserved 750 million, 1 billion, or even 2 billion years ago are essentially indistinguishable from living forms. This is quite different from the fossil records of plants and animals, which are replete with extinct forms.

And here is further food for thought (Magulis & Sagan, 1997, p. 72):

> Barghoorn's Swaziland discovery of actual 3,400-million-

> year-old fossil microbes raises a startling point: the transition from inanimate matter to bacteria took less time than the transition from bacteria to large, familiar organisms.

For three *billion* years there was nothing but bacteria on the planet, then approximately 600 million years ago . . . boom! Complex life forms appear (Raup, 1992).

But the question remains, if an organism does not evolve after a century, or a thousand years, or ten thousand years, or a million years, just when is it supposed to evolve? And if evolution does not take place after ten million (sand dollar), twenty million (Voltinia butterfly), fifty million (purple frog), one hundred million (coelacanth), one hundred and fifty million (Chimaera fish), two hundred million years (Bangia algae, brittle starfish, sea urchin), *then when, exactly, is it supposed to take place?* After all, there is a finite number of millions of years before we enter the era when the Earth was a ball of molten matter.

The Cambrian Fossils.

The first thing that strikes one when first seeing Cambrian fossils is that they are bizarre. They are just bizarre looking, so unlike anything that is alive today, though we do see occasionally the beginning tendency towards cylindrical morphology (Chen *et. al.*, 1995; Parker, 2003; Conway Morris, 1998) beginning to take place due to the advent of the notochord. With some of the organisms, it is impossible to say which is the head and which is the tail---assuming that there is a head---which is up and which is down. In fact, one of the fossils, aptly named *Hallucigenia,* was at one time pictured upside down---assuming, of course, that now we have it right (Gould, 2007). The Cambrian fossils are much, much, much older than the ones from the Jurassic or the Eocene.[10]

And this is the point that I want to make: if evolution was, indeed, going on constantly, inexorably, through Natural Selection, then not a single fossil---from any era whatsoever, be it the Miocene, or the Jurassic---would be recognizable. It certainly would not be *identical* to any living species today. There should, furthermore, be no coelacanth, no sea urchins, no starfish, no stingray, no shrimp, no sand dollars, no dragonflies, no gnats, no crayfish, no rays, no cockroaches, no ants. And if they were alive then, then they should not be alive today. *All* fossils should be simply unrecognizable, totally alien, and the farther back that we go, the more alien and bizarre should be the morphology. Yet, the only instance where this is the case is the Cambrian period.

One of the early criticisms of the Darwin-Wallace theory that was considered to be simple minded may not have been so after all: "If evolution is going on all the time, why don't we see it taking place?" The answer that has always been given is that it is taking place, but it is such a long process that it takes centuries, or millennia, to see the results.[11] Yet, that only makes sense if evolution was beginning now; if evolution is constantly taking place, there should be at least one end result taking place now.

Lastly, let me point out that the existence of living fossils was known even before the 1800s, although the term "living fossil" seems to have come later. Wallace (1881/1998): "In the Pliocene and

Miocene formations most of the land shells are either identical with living species or closely allied to them; while even in the Eocene almost all are of living genera, and one British Eocene even lives in Texas." (p. 74) And Charles Lyell (1833/1997): "Let us suppose those hills of the Italian peninsula and of Sicily, which are of comparatively recent origin, and contain many fossil shells identical with living species, to subside again into the sea . . . ." (p.61)

Punctuated Equilibrium. Being an outsider to the field, it was after I had begun to work in earnest on this project that I became aware of the punctuated evolution contribution of Eldredge and Gould (1972). The two paleontologists had thoroughly examined the fossil record of gastropods and trilobites and realized that gradual evolution was not taking place as classical theory predicted. Instead, there were long periods of stasis, with no evolution taking place, followed by bursts of evolutionary activity (Gould, 1995; Broyles, 2004). Along with Bakker (1986), punctuated evolution gave the stagnant field a much needed kick in the pants (Prothero, 1992). Opposing camps inevitably coalesced (e.g., Gould vs. Dawkins), with the attacks from neo-Darwinists being their usual viciousness (Gould, 2007b), but opposition diminished once Gould and Eldredge consistently made it clear that they were not trying to bring down Darwinism, but were buttressing it (Gould & Eldredge, 1993); indeed, they extended the life of Darwinism with this epicycle, but only if other scientists had started looking at the data the same way that they did. At any rate, once it was pointed out, scientists have been steadily confirming punctuated evolution from the fossil record---again, once it was pointed out (Adler & Carey, 1980; Williamson, 1981; Kelley, 1983; Prothero, 1992). One team even carried out experiments on the topic (Elena, Cooper, Lenski, 1996; Mlot, 1996; Coyne & Charlesworth, 1996; Pagel, Venditti & Meade, 2006), although others have cast doubt on punctuated evolution in certain cases (Kerr, 1996). Inexplicably, to my knowledge, no one has pointed to the more obvious, much more dramatic evidence of living fossils, which fact even Mayr called "puzzling," clearly an understatement (Mayr, 2001).

(what I find particularly interesting from a psychological perspective is that Gould (2007) states that once the furor died down, paleontologists realized that punctuated evolution was a fact that had been staring at their faces for decades and they saw evidence of it everywhere, e.g., Brett & Baird, 1995).

Of related interest, mathematical models that have tried to replicate evolution have invariably ultimately achieved equilibrium and, hence, *have been deemed defective* (Williams, 1996). It may very well be that *they replicated evolution all too well.*

It is telling that in their original presentation of their concept of punctuated equilibrium, Eldredge and Gould (1972) did not see fit to mention or cite two German paleontologists who had, independently, been *the first* to have observed and reported the same findings (i.e., *they had precedence*), Rudolf Kaufmann (Fortey, 2000) in examining trilobites and Otto Schindewolf (1950/1994) in examining ammonites, although Gould later did condescend to write an introduction to Schindewolf's book. Schindewolf (1950/1994), who did exhaustive research on ammonites and corals, had come to similar conclusions. What is particularly important to realize here is that whereas adherents to the classical theory have maintained that no intermediary forms between species have not been preserved, though they must have existed, because fossils are so rare (that is, vertebrate fossils), that is not the case with ammonites and trilobites; figuratively speaking, one could drown in ammonites, trilobites and mollusks, they are that plentiful. Schindewolf is worth quoting at length at this point (p.214; italics in the original):

> According to Darwin's theory, evolution takes place exclusively by way of slow, continuous formation and modification of species: the progressive addition of ever newer differences at the species level results in increasing divergence and leads to the formation of general, families, and higher taxonomic and phylogenetic units.
>
> Our experience, gained from the observation of fossil material, directly contradicts this interpretation. We have found that the organizing structure of a family or an order did not arise as the result of continuous modification in a long chain of species, but rather by means of a sudden, discontinuous direct refashioning of the type complex from family to family, from order to order, from class to class. The characters that account for the distinctions among species are completely different from those that distinguish one type from another.

Unfortunately, in America if a report is not written in English, it does not exist (cf. Reif, 1986).

Other paleontologists, A. A. Hallam, studying mollusks, (Schwartz, 1999) and Kelley (1983) also studying mollusks had also come to the same conclusion---again, based on *data*.

Although Eldredge and Gould shook up the field with their proposal of punctuated evolution, they were both too timid to follow it up to its logical conclusion, namely that the classical theory was flawed. One cannot really blame them for not doing so. In a way, it would have meant cutting the ground from under their own feet and would have, in their thinking, have resulted in the triumph of Creationism (Eldredge, 2001). Yet, even so, had they taken the next logical step, considering how their more modest article affected their colleagues, it would have truly revolutionized the field.

But, then again, if Natural Selection is not the driving force behind speciation, what is?

Convergent Evolution. A shortcoming of science that occurs fairly frequently is a psychological one in that whenever a curious phenomenon is encountered and a name is given to that phenomenon, by an odd psychological consequence, an emotion occurs that gives one the impression that the phenomenon has been explained in its entirety simply by naming it. For example, when the question is asked, why does an animal emit a particular, complex, behavior, such as a pigeon finding its way home when released hundreds of miles away, the "answer" is given: instinct---it is instinctual. Why does the spider spin a web? Instinct. Why does a cat lick its newly born kittens? Instinct. A feeling of satisfaction ensues and further inquiry seems to become suspended and the how and why that would have naturally arisen is ignored for a long time thereafter.

Such is the case with so-called "convergent evolution."

One of the more fascinating aspects of the question of evolution is the topic of convergent evolution. Convergent evolution occurs when two totally different organisms have an identical, unique, trait which they both evolved independent from each other. The way that it is usually presented is that when faced with a "problem," two organisms will evolve in the same way in order to challenge the obstacle. When presented thusly, one cannot help but again imagine a couple of animals poring over blueprints, arguing and coming to the conclusion that the trait in which they will share is the only viable solution to the "problem." Even such a neo-Darwinist as Dawkins (1982, p.65) succumbs to this: "But given the admitted advantage of synchrony, why didn't the cicadas settle on a shorter life cycle than 17 (or 13) years, thereby reducing the unfortunate delay in reproduction?"

And also (1986, p.85):

> Why, in all the hundreds of millions of years since its [Nautilus's] ancestors first evolved a pinhole eye, did it never discover the principle of the lens? The advantage of a lens is that it allows the image to be both sharp *and* bright. What is worrying about *Nautilus* is that the quality of its retina suggests that it would really benefit, greatly and immediately, from a lens.

As well as Ernst Mayr (2001): "Certain deep-sea fish have dwarf males that are attached to the females, because free-swimming males might have difficulty finding females in these vast and lifeless spaces." (p.139) And, an anonymous editor, or writer, from *Nature:* "Flowers of the neotropical vine *Mucuna holtonii* have evolved an intriguing way of making sure that their pollinators don't make wasted journeys to flowers not yet ready to release the pollen." (Anonymous, 1999, p. ix) Lastly, another famous neo-Darwinist, Julian Huxley (1953):

> Color vision in one organism generates color in others. Flowers develop distinctive colors [in order] to attract bees; wasps develop their yellow and black [in order] to warn enemies of their stings; the partridge develops camouflage [in order] to escape detection by the hawk; the peacock develops brilliant plumage [in order] to stimulate his mate. (p.79)

But, to get back to convergent evolution itself:

A superb example of convergent evolution are, again, pitcher plants. Pitcher plants occur in the continents of North America, South America, Australia and (southeast) Asia. Europe and Africa are devoid of pitcher plants, though the island of Madagascar does have them---for the time being.[11] Human beings being the bipedal locusts that they are, several species will become extinct in Madagascar before too long, along with the lemurs.

The various species of *Sarracenia* pitcher plants are located on the eastern part of North America. In the western part, we find only one species, the Cobra Lily, *Darlingtonia californica*. It follows pretty much the same overall pattern as the *Sarracenia*, with the exception of two structures, a dome over the top of the leaf and a fishtail-like appendage near the opening. One other difference is that the Cobra Lily does not produce enzymes; bacterial action breaks down the trapped insects. Otherwise, it is pretty much the same as the *Sarracenias*.

Let us now pass on to South America. The only pitcher plants in the entire continent are the four species of the genus *Heliamphora*, which are found in the northeast, in Venezuela and Guyana plateaus (Lloyd, 1942). Again, we see the same pattern, though they, too, produce no enzymes.

Next, we go to Australia, where we find only one species, *Cephalotus follicularis*. Again, the same pattern and *Cephalotus* does produce enzymes.

When we come to the *Nepenthes* genus, things get a little bit more interesting. The traps themselves, though, follow the same predictable pattern of development. There are 95 species of *Nepenthes*, with naturally occurring hybrids, and the variation of pitchers' shapes, colors and sizes are fascinating. Their distribution centers in Indonesia, though they are known in the Philippines, Ceylon, Papua New Guinea, Australia, Malaysia, eastern India (some species, like *Nepenthes khasiana* are now extinct in their natural locations) and Madagascar; it was in Madagascar[12] that Westerners first encountered and described *Nepenthes*. Other characteristics that differ it from other pitcher plants is, first, they are not self-pollinators, they are, instead, dioecius, and second, the overall plants are vines (Lloyd, 1942; Schwartz, 1974).

Now, then, orthodoxy dictates that, in convergent evolution, two species, independent of each other, evolved a multitude of traits (actually, it is always put forth simplistically as *a* trait) which were adaptive. With pitcher plants, we have convergent evolution working overtime, for we have convergent evolution taking place not just in two species, but in five genera. There is no proof---no proof whatsoever!---none!---through experimentation, DNA analysis, or cladistics, that the pitcher plants in South America, North America, Australia, Asia and Madagascar (the genus *Nepenthes, Cephalotus, Sarracenia, Darlingtonia* and *Heliamphora*---and each and every one of their respective species) all formed their pitchers through a mind boggling sequence of cumulative traits---each of which, by itself, had a dubious adaptive value---through "convergent evolution," i.e., through Natural Selection. If one was a statistician with plenty of time on his/her hands, it would be interesting to calculate the probabilities involved in such a phenomenon---but no matter what the outcome, the neo-Darwinists would not be fazed. It is a matter of *belief*, pure and simple, belief in Natural Selection rather than some other, unknown, mechanism. A similarity of complex morphology is present across five genera and by giving the similarity a name ("convergent evolution") the illusion is established that the name itself is the explanation for the phenomenon. Again . . . note that *no proof, no data, is offered* that indicates that evolution took the organisms through the same sequence.

A name is given: "convergent evolution." And that is supposed to explain everything.

The idea, however, is intuitively absurd. Nature does not reinvent the wheel with each species (Soria-Carrasco, 2014).

However, let us focus, instead, on an instance of "convergent evolution," again on the very same pitcher plants, but of a *nonadaptive* trait which all five genera share, along with their adaptive predation. I am talking about the flap that is present, in one degree or another, in all the genera.

In *Sarracenia minor, Sarracenia psittacina* and *Darlingtonia californica,* the flap is actually a dome that encompasses the opening of the trap; in both, the dome does serve as a beneficial trait in that flying insects can be seen bumping against one of the white spots on the dome, trying to get out. In all the rest of the pitcher plants, however, the flap simply protrudes over the opening, and, of significance, in *Heliamphora,* it is just a vestige of a flap, similar to the coccyx being a tail in humans. The flap serves no function whatsoever. It does not prevent insects from flying away once they begin to descend because they rarely try to do so. It does not prevent rain from entering the pitcher during a rainstorm (*Nepenthes* in the tropics is subjected to downpours), especially rain driven by the wind. If anything, the presence of water in the pitchers helps to break down nutrients.

So . . . why *did* the flap evolve into being---in all five genera---if it has no adaptive value? I think that the answer is obvious: it did not. That is, it did not evolve separately. The flap is part and parcel of pitchers. If a plant develops pitchers, it will develop flaps; it is a package deal, sort to speak. In other words, there may be "genotype clusters" whose basic components are indivisible and make up pitchers. In the case of pitcher plants, these basic components show up in each genera, but each species can vary from each other with respect to other incidental traits, e.g., coloration, size of pitchers, flowers, shape of rim and flap. Even within a species, one encounters subspecies with uniform traits; for example, in *Sarracenia purpurea,* the pitcher in northern latitudes is larger and colored maroon with venation while the flower is likewise maroon; in southern latitudes, the pitcher is smaller, green and the flower a yellow-green in color.

The idea of "genotype clusters" (wherein a cluster of traits are linked, only one of which may be said to be truly beneficially adaptive, if at all) makes more sense in explaining "convergent evolution" than to theorize that Natural Selection slowly selected for each individual component of each individual pitcher plant species on the basis of the adaptability benefit to that individual plant of that individual trait---across five genera.[13]

Another example: the saber tooth tigers, like *Smilodon,* along with their marsupial counterparts of South America, have their famously protruding canines. Those huge canines are usually thought to be used by the carnivores as stabbing implements. I seriously doubt that---if for no other reason than so many are found broken---and feel that they were maladaptive and got in the way. The fact is that those two cases of protruding fangs are pointed to as being a case of convergent evolution through Natural Selection, but, the similarity (and the "explanation") usually ends there. Yet, in point of fact, we see that very same exaggerated dentition in previously extinct species: *Aulacephalodon, Dinodontosaurus, Bathyopsis,* and *Uintatherium*---and *all of them are herbivores*. Three other saber-toothed herbivores, *Dicynodon, Rhychosaur, Lystrosaurus,* date as far as the Permian era.[14] And we see it today in species such as the elephant and babirusa.

Likewise, we see the beak surfacing from time to time: in parrots, in squids, in ceratopsians.

And, we see that the four-chambered heart developed independently in birds and mammals (Denton, 2016)

In short, let us use Occam's razor.

One Species, Two Species, Three Species? Plants are rarely mentioned in discussions of evolution. Usually, the focus is on animals. Anyway, in my recent collecting trip, I obtained in one spot half a dozen sundews, genus *Drosera*. The problem, though, is that, without their flowering, I have no idea what species they might be. When I used to collect another sundew species, *Drosera filiformis,* in Mississippi before they were all crushed to death under the developers' bulldozers, there was no mistaking it. *D. filiformis*'s leaves slowly uncoil upwards to make a superb barrier to mosquitoes, if one happens to be lucky enough to have a lawn full of these wonderful plants.

But the ones that I gathered on this trip are not so distinguishable. They could be *Drosera rotundifolia,* of course. Or they could be *Drosera anglica*, or, for that matter, they could be *Drosera intermedia* or, again, *Drosera capillaris*. I just cannot tell them apart (incidentally, this is a very common occurrence in botanical organisms, still another basic difference between flora and fauna).

And this, precisely, is the point. If these four species are practically identical with each other---and this is not such an unusual occurrence in Nature: in fact, I have the very same problem with the butterworts

(Pinguicula) that I collected on the same spot---how can Natural Selection have selected for each one? On what basis did Natural Selection differentiate each other to such an extent that they formed four different species? That is, if one adheres to the classical theory.

One could even muddy up the waters further by stating that nearly identical species within the same genus are a form of convergent evolution.

Other Flaws with the Classical Theory

Cats and dogs. Darwin used selective breeding of plants, pets and livestock by humans as an analogy to Natural Selection; additionally, he stated that artificial human selective breeding greatly accelerated the process of Natural Selection so that, instead of centuries, variations could be obtained after a few dozen generations. It would be much simpler if I were to choose cats for my example since their appearance, in spite of centuries of selective breeding, have not really changed all that much (we have not seen a cat the size of a Great Dane, for example). However, let me purposely choose dogs because of their wide-ranging appearance of which most people are familiar with.

My point is this: with dogs you have the genotype, which dictates what is a dog. Otherwise, it is not a dog. And within those parameters---and especially with dogs---you have a wide latitude of potential variation. But you can go no farther. With dogs, unlike with cats---and this is significant of itself---you have a wide range of morphology: you have the repulsive, rat-like Chihuahuas, you have the carnivorous horse-like Great Danes, you have the prune-like sharpeis, the hotdog-like dachshunds. *But, you still have a dog.* Yes, millennia ago, the dog evolved from the wolf (without even one chromosomal change (Stindl, 2014)). But, since then, the point is that it does not matter how many variations you get, no matter how often, or how long, you will carry out selective breeding, you will *always* get a dog. Not a cat. Not a bird. Not a ferret. A dog. Never a non-dog. There will be no speciation. Try as you might, no matter how many centuries you apply yourself exclusively to selective breeding, you will never turn a dog into a cat. Or a badger. Or a rabbit. And for that matter, you will never be able to turn a rose bush into a petunia.[15]

And another very important point: experienced plant breeders and animal breeders know that if all the different varieties of organisms that they have selectively bred for over decades were allowed to breed at will, the end result will be the original "mutt." Backcrossing.

From time to time we hear or read a story wherein lethal bacteria have adapted and acquired immunity to certain antibiotics, whereupon it is concluded that the bacteria have "evolved" (Huxley, 1953; Trivedi, 2001; Gilliver, Bennett, Begon, Hazel & Hart, 1999; Palumbi, 2001). Likewise, a story may come our way about fish being moved to another stream with a stronger current, whereupon its descendants grew bigger fins with the same conclusion being made (Hendry, Wenburg, Bentzen, Volk & Quinn, 2000). Or, that an island plant's morphology has slightly altered from the mainland conspecific (Diamond, 1996). Or, of fruit flies which have been introduced into another continent, whereupon their wings changed a little (Morris, 2001). Or, that the same species of salmon are somewhat different in size and coloration depending on whether they are in beaches or rivers (Hendry, 2000). Or that guppies in different streams differ in body and litter size (Akst, 2017). However, the fact of the matter is that no speciation has actually taken place. What you have is the same fish, but with bigger fins, the same fruit fly but with altered wings, the same plant but with a small achene and the same bacteria but with a new resistance to chemicals (Weiner, 2005). Another recent study (Losos, Shoener, Langerhans & Spiller (2006)), an experimental one for a change, claiming that their findings proves evolutionary change through Natural Selection, has reported that, upon introducing a predatory species of lizard, another species of lizards has developed longer legs and has become more arboreal (it proves nothing regarding *speciation*. Again: humans have for centuries proven that a species' morphology can be somewhat altered through selective breeding. It is nothing new. It is not evolution, it is not speciation).

These slight alterations of animals' morphology in the field due to environmental changes continues to be trumpeted each time as if they were a major discovery, but human beings have been doing the very same thing for centuries with crops and animals and they continue to do so, but in neither case has speciation been the end result. Additionally, if a slight change in morphology does take place, catalyzed by severe environmental alterations, such as finches' beaks in the Galápagos during periods of severe drought (Weiner, 1994), it remains to be seen whether a return to the mean occurs once the temporary period of the altered environment ends.

And *apropos* of climate change, we have various organisms which, for the past forty years have been slowly adopting to climate change through alteration of their behavior or their range of habitat (Robbins, 2004; Bradshaw & Holzapfel, 2006). No speciation has occurred.

In short: *there has been no speciation through Natural Selection.*

Successful Adaptation. Up until the mid-1970s, it was popularly believed that dinosaurs died out because they were big and stupid and they had to make way for us. Now, of course, it is widely acknowledged that dinosaurs were exceptionally well adapted to their world, which is why they ruled it for so many millions of years and they would still be around, if not for a fluke.

And this brings us back to the subject of bacteria. Bacteria are ubiquitous. They have been around since the beginning of life. They are extremely successful and hardy, so why did they have to evolve at all into "higher organisms," anyway? Sponges have been around since the Cambrian, almost certainly since the Ediacara. Somewhat recently, archaea bacteria were found in 3.5 billion years old rocks (Furnes, Banerjee, Muehlenbachs, Staudigel & de Wit, 2004; Raup, 1992; Knoll, 2003). Why did they also *have* to evolve into so-called "higher organisms"?

Better yet, stromatolites. Stromatolites are the living fossils of all living fossils. *For 90% of Earth's existence*, stromatolites ruled the Earth. 90%! They existed not for millions of years, like dinosaurs, but billions of years. As Walker (2003) rhetorically asks, why should they have evolved at all?

Individual or Group Evolution? By now, the reader is aware of the litany on how Natural Selection brings about speciation. If we adhere to classical theory then the threshold into speciation is all important.

An organism suddenly develops a trait (say, white coloration like the lions of Timbavati (McBride, 1977)) that happens to be advantageous. Fine. It---not "they"---has more offsprings. Fine. They now all have that advantageous trait; they are all white. Fine. But, you *still* have the same organism---except that it is now white. And you will always have the possibility of backcrossing with other members of the original population.

Here is a different scenario. An organism develops that same adaptive trait, but it is physically isolated from the main population. Fine. The isolated organisms are now white, whereas the parent population is brown. Well and good. And just because we feel extra generous this morning, we will bestow on the isolated population, for good measure, five more advantageous traits. Good. But . . . it is *still* the same species. The *only* way that it could become a different species is if the sexuality itself was directly affected. That is, if the isolated species, instead of wanting to mate when both the male and female did a little dance together with their necks bent backwards now, suddenly, inexplicably, wanted to mate by abandoning the dance altogether in favor of the males building an earthen mound. And, anyway, how on earth could that possibly be more adaptive, more selected for?

Zahavi and Zahavi (1997), although not studying insectivorous plants, are equally puzzled by orthodoxy:

Feathers, like horns and antlers, present a puzzle to researchers of evolution. Birds clearly evolved from reptiles and feathers from scales. But it is unlikely that one mutation created something as complex and as beautifully functional as a feather. Nevertheless, in order for any change to spread in a population, it must be an improvement; one can readily see how a body member that has a particular function or purpose can evolve to better serve that purpose, but how can a body member like a reptilian scale evolve gradually into a totally different one like a feather? The transformation of a scale into a feather can only have occurred through a series of countless tiny changes, one after another; each of these changes could spread among the population only if that change, on its own merit, enhanced the fitness of an animal that carried it--that is, the number of reproducing offspring that it had. Yet obviously, the changed scale became a less efficient scale long before it turned into an efficient feather. (pp. 90-91)

Classical theory as it is conceived (Dawkins, 1995) is erroneous in implying that it is an individual (A12) who acquires the last, capstone, trait (out of, say, twelve traits) that results in speciation. *And where is it going to find a mate?* If it attempts to mate with his capstone-less conspecifics (A11), then it (A12) does not constitute a different species from them (A11). By the same token, if they (A11) could mate with their immediate predecessors (A10), the whole process could have gone back all the way to the very beginning (A0). Remember that when varieties of domestic animals and plants are allowed to interbreed, they revert to the A0 generation, the "mutts."

However, by the time we reach the capstone generation (A12), that one individual organism is so unique, both genetically and behaviorally, that it cannot mate successfully and, therefore, the capstone trait dies off. Two organisms acquiring the capstone trait simultaneously, one a male and the other a female, through the mechanism of Natural Selection, is extremely improbable, if not impossible. Additionally, the genetic variation of those two lone organisms would be so narrow as to constrain later survivability in their offsprings.

Then, again, there is the fatal *caveat* of backcrossing to take into account. Consider, for example, the recent declaration by some (Holland, 2007) that the disappearance in the human species of blondes and redheads is inevitable since they are fewer and are vastly outnumbered by brunettes, even though blondes and redheads are seen as more attractive and therefore should be selected for according to the classical theory.

Rather, I propose that when speciation does take place, it takes place on a group basis rather than on an individual basis. In the next chapter we will explore this further.

Another detail: plants and animals show one other important difference, other than the usual ones. With animals, in the natural world, interbreeding of species is practically unheard of.[16] In the plant world, it is not uncommon. *Sarracenia* and *Nepenthes* hybrids, for example, are occasionally found in the wild.[17]

Another factor: adherents of the classical theory love isolated populations, particularly in islands. Thus, to them, this isolation facilitates speciation and prevents the dilution of any advantageous traits into the larger population (MacArthur & Wilson, 1967/2001; Weiner, 1994). Some neo-Darwinists would even insist that physical isolation is a *sine qua non* for speciation (some biologists disagree (Pennisi, 2014)). And yet, we find, first, that some species which have been physically separated by two continents not only to be similar, but interbreed with each other with no problem whatsoever, like the skunk cabbage. Secondly, even though there has been no physical isolation, speciation does take place in Nature, in defiance of the official theory (Morell, 1996, 1996; Gibbons, 1996, 1996; Tregenza & Butlin, 1999; Morris, 2001; Mayr, 1992, 2001; Meyer, 2004). Losos and Schluter's (2000) study with the *Anolis* lizards of the Caribbean illustrate this very well. On the large islands of Cuba and Hispaniola, with geographical diversity, we do, indeed, find a large number of species; yet, on another large island, Puerto Rico, we do not, just as we do not find it in the larger small islands which also have a geographical diversity. Also, away from the Caribbean, in the very large and geographically diverse islands of Sri Lanka and Taiwan, there is little diversity in the number of lizard species. And the lakes where so many species of cichlids have come into being (in an incredibly short time) are nearly uniform in geography and composition.

Zahavi and Zahavi (1997) make an additional, relevant, point:

> How, then, does ritualization work? How do movement signals evolve from movements that used to have another purpose?

> A signal's value to the signaler is that it can convey information to another individual. But the message can be conveyed only if the other individual is interested in the message and understands it. The process therefore cannot start with a mutation in the signaler; because that would require *two* simultaneous, coordinated mutations: one which caused the signaler to perform the signal, and another causing the observer to take interest in it and understand its meaning. Even in the highly unlikely event of two such simultaneous mutations, the chance of the two mutants meeting is practically zero. (p. 66)

Dollo's Law. This Belgian law simply states that evolution is irreversible.

> For this reason, one persistent principle governing the process of evolution---one could almost say the only principle to which there is no exception---is all the more important. This is Dollo's Law of the *irreversibility of adaptation.* Although it has no theoretical basis at present, it is an empirically binding rule that phylogenetic processes leading to the differentiation of characters and organs *are never reversed to lead back to the original condition.* (Lorenz, 1996; p. 128; italics in the original).

This "law" has to be false. No one would disagree with the statement that flight is an enormous advantage to an organism for any number of reasons. Yet, we see that certain species of birds took a step backward and became flightless: kiwis, emus, ostrich, cassowaries, dodos, moas, rheas. Note, first, that size was irrelevant and, second, the abandonment of flight resulted in the extinction of the dodo and the moa, followed by the present near extinction of the kiwi.

An additional argument involves the blind cave fish and insects among troglobites, those famous organisms who live in total, uninterrupted darkness (Krajick, 2007). The curious thing here is not that they are blind, but in that their eyes have *disappeared*. That complex, adaptive organ which gave Darwin so many headaches in trying to imagine its formation through Natural Selection---gone. It is one thing to claim that sightlessness is present in cave fishes in caves where you have eternal darkness, due to atrophy of the optic nerve (Espinasa & Espinasa, 2005); it is a whole kettle of fish to explain *eyelessness*.

Another example: a protozoon named *Entamoeba histolytica* is an anaerobic parasite, yet genetic analysis reveals that it once had mitochondria (Knoll, 2003).

Biologists tend to avoid explaining when and why evolution turned the clock back. No one can possibly deny that the acquisition of flight is a boon to an organism. It is an evolutionary milestone. Many of the dinosaur species flew (Lessem, 1992). Birds, supposedly a distant cousin of dinosaurs (Bakker, 1986), have enjoyed the advantages of flight. Countless times I have witnessed housecats involved in their futile attempt to catch a bird, which flew harmlessly away, leaving the frustrated feline predator behind. What, then, can possibly be the advantage of an atavism like flightless birds? Not just in the diminutive kiwi, or the dodo, or the Pleistocene flightless Barn Owl from Cuba (Alcorn, 1986). What possible advantage can it be to go back to being flightless? None! If anything, it is a very obvious disadvantage. It is *maladaptive*. Yet, no less a figure than

Mayr claims that evolution can be retrogressive, not realizing that the statement is an oxymoron (Mayr, 2001).

Flightlessness was not selected for as being more adaptive through Natural Selection. Rather, flightlessness was *imposed* upon them (and others along the way which have also become extinct).

And speaking of the dodo and the moa, let me point out that they did not evolve once it came time to do so, i.e., when they were under attack. They simply became extinct. We see that the same thing happened in the Hawaiian Islands when cuddly, fuzzy, predators were introduced by humans into the ecosystems. One of Darwin's finches soon became extinct when it was used as a source of food by humans.

Near Tupelo, Mississippi, as one drives northward, one can see the pleasant verdure to either side, which tends to evoke a pleasant feeling until one realizes that the entire landscape has been taken over and covered over by one organism, the same organism, the kudzu, an aggressive nonnative plant (neither the state government nor the federal government has seen fit to stem the invasion because politicians). The whole point here is to highlight the fact that when an organism is faced with a new, aggressive, efficient predator (or competitor), it does not evolve---it dies. The organism has no time to "study" the "problem" and to adapt to the "problem" and then evolve. It just dies. There is no time. Death comes too quickly. It becomes extinct. Just look at Guam. Or the Hawaiian Islands. Or for that matter, look at the humans originally dwelling in Cuba, Hispaniola or Jamaica, known as the Taínos, the Caribs and the

Siboneys; when the Spaniards arrived and became the predators, the Taínos, Caribs and Siboneys were wiped out. None remain to this day.

Ever since the introduction of nonnative plants and animals into Australia, New Zealand, America, Britain, Cuba, Hawaii, St. Helena, Guam, Mauritius and other islands, not one instance of evolution through Natural Selection, i.e., speciation, has occurred by the native organisms in order to deal with "the problem." Not one. Bring this fact up to a neo-Darwinian and he/she will patiently---and condescendingly, always condescendingly, and with a sigh---explain to you that evolution takes centuries, millennia, millions, tens of millions, hundreds of millions of years to take place---then sit back as he enthusiastically tells you, with a gleam in his eye, of instances where fish or flies were introduced into a new environment and their descendants' wings or fins got bigger, in only a couple of generations, clear evidence of evolution at work, (incidentally, the rapid slight changes in morphology of animals put in a new environment was noted by Wallace (1881/1998) who did not remark on the apparent contradiction of the theory).

Other objections, and exceptions, to Dollo's Law can be read in Collin and Cipriano (2003) as well as Reebs (2004).

Random Death. Most of us in Western societies have become so accustomed to our civilized lives with its steady, uninterrupted flow of food into the overpopulated cities, that we have little inkling as to how truly harsh life is in the wild---particularly in the wintertime with starvation being a very real, every day, possibility. Certain of the environmentalists, however, have this absurd illusion of the wilderness being like a well-behaved, well-kept, English garden that Westerners have corrupted, and one occasionally sees them blithely ignoring warnings when in the presence of predators, with predictable results (of peripheral humor, I am always amused at the sight of Europeans who, having virtually exterminated the wildlife of Europe, travel thousands of miles to South America, Africa and Asia to lecture its inhabitants on wildlife conservation).

Having said that, let us travel back in time to the Stone Age: I have just suddenly acquired a mutation, conveniently by the snap of a finger, whereby all my offsprings will be twins (and, furthermore, I do not live with the Kikuyus who have traditionally killed twin babies). Great! Unfortunately, that year there was a drought and a forest fire which killed my first mate in her eighth month of pregnancy. My second mate died of starvation that following winter. My third mate died during childbirth, along with the babies. My fourth mate fared much better, but one of the twins died when it was suffocated by its umbilical cord. My other twin child was seized and eaten by a predator. My second set of twins lived to the age of five intact, whereupon both died of diphtheria. My third set of twins lived up to adulthood, at which point they and I were killed and eaten by another band of cannibalistic humans, which is just as well because, soon after that, three quarters of my own tribe died of the smallpox.

Anyone who is acquainted with pre-civilized societies prior to the advent of antibiotics and agriculture will admit that my gloomy scenario is not too exaggerated---and I left out the parasites. The point however, of my depressing tale is to point out that classical theory does not take into account the random, frequent death that occurs in Nature.

Apropos of this, one of the questions plaguing ecology, biodiversity and biogeography involves the number of individuals left of an endangered species. It is a truism that the larger the number of individuals in a species the less likelihood that it will face extinction; the inverse, of course, is that the scarcer the species is the more likelihood that it will become extinct when faced with pathogens and/or temporary climactic alterations and/or genetic aberrations. There are some species whose members nowadays number less than a mere fifty, the threshold number for some conservationists to consider such species as a whole as a hopeless lost cause and, instead, focus limited conservation efforts on other, more numerous, species that are potentially salvageable (other conservationists, by the way, become upset at this fatalistic triage (Quammen, 1996)). Others argue that the viability threshold number is different for different species. Regardless of the specific number and whether or not one should adopt a fatalistic outlook, there are several well documented examples in nature wherein a rare species ultimately became extinct because of either a) demographic stochasticity---accidental variations in the birth rate or death rate, age of surviving members and in the ratio of sexes b) environmental stochasticity---fluctuations in the occurrence of pathogens, the food supply, the number of predators (or prey) and weather conditions c) genetic stochasticity---the accumulation of harmful, recessive alleles, or, conversely, the disappearance of beneficial alleles, or, the loss of genetic variation. Or a simultaneous combination of all three in a relatively short period of time, not at all an unusual occurrence. In short, small populations of an organism are statistically vulnerable to extinction.

These concerns have become crystallized in the past two or three decades because of pollution, global warming, poaching, etc. They were certainly unknown at the time of Wallace and Darwin. Nowadays, neo-Darwinists smugly uphold the classical view in isolation of the above concerns. Yet, the above problems directly affect the subject of the evolution of species. Apart from other questions (such as backcrossing), the genesis of new species is traditionally seen as having few members, i.e., the species, though superior in some way and in some degree to its former conspecifics, is a rare, endangered species (Boessenkool, et al., (2007). And even if we do not contemplate speciation, but, rather, a brand new beneficial trait within an incipient species (such as the white lions of Timbavati), the same Damocles sword hangs over their heads.

Human Races. I will touch, very briefly, on the question of human races (Coon, 1965; Balter, 2006). Within the Sub-Sahara Africa there are a number of "races," or, if you prefer, "ethnic groups," or, if you even prefer, "subspecies," whose physical differences are very obvious and very striking. The Bushmen are of short stature (5' tall) and of slender built, with yellowish skin. Their faces are flattened with slanted eyes, almost no body hair and peculiar genitalia (Marks, 1991). The Pygmies are even shorter (3'), of reddish skin, bulbous foreheads and protruding eyes. The Masai look as if a child giant had taken a human made of rubber and pulled until it could no longer do so. They are extremely tall and thin.

Three points: First, there is not anything within all of their respective physical attributes that could be said to have been particularly adaptive, so that they were selected for through Natural Selection and become fixed. Second, it is *a group* that was affected for those particular traits, not individuals: short stature in males is a trait that---very strongly!---goes counter to Sexual Selection in every, or almost every, human culture. Third, it would be of the utmost importance to be able to determine just when those groups came into being.[18]

Sex with Humans. When Desmond Morris (1967) published his theory of human sexuality, it caused a great stir at the time, but with the advent of the feminist movement it was suppressed as being Politically Incorrect. His Naked Ape theory states that sexuality plays a more hedonistic, more important role in humans than is usual in the animal kingdom and that it is this hedonism that leads to bonding between partners. Part of the human anatomy has been modified for this purpose. Human females have developed breasts and their torsos are hairless while males have developed a large penis and both genders can experience orgasms. This particularly leads to front-to-front mating which, in turn, promotes bonding.

Regardless as to whether or not one adheres to the Naked Ape theory, most everyone would agree that (a) sex does play a principal role in humans' lives and (b) large breasts in women are very appealing to males.

Consequently, classical Darwin-Wallace theory

would *predict* that women having larger breasts would have an adaptive advantage over other women in that they would attract more mates and/or would have more offsprings. Through time, all women would have large breasts while flat-chested women would fall by the wayside and become extinct. Humans have been around for a long, long time, time enough, if classical theory was valid, for every woman walking around nowadays to look like Dolly Parton. Sadly, this is not the case.

And why not? Well, because we see that other factors come into play and because no matter how marginal in physical appearance or social status, every human, male and female, ultimately finds a mate.[19]

The crucial point here, though, is that the classical theory's *predictability* fails. And one of the prerequisites for any theory to be a valid one is that it must have predictability and, therefore, can be tested.

Cope's Rule

Of related interest: The nineteenth century colorful paleontologist, Edward Cope (Jaffe 2000), asserted that the longer a species evolves, the bigger it gets. The favorite example that he had in mind was the horse. Although the rule was questioned from a theoretical basis, it was not until the end of the 20th century that it was actually tested empirically from the fossil record; like so many assertions about evolution, nobody had ever gotten around to actually empirically testing this proposition. At any rate, it was found that many lineages shrank as well as expanded (Santos, 1997).

Darwin vs. Dawkins

Darwin was obviously a brilliant scientist, no

question about it. So was Wallace. Both hand minds of the first caliber---even without taking into account their joint theory. Let me repeat that: even if you exclude their famous theory, the rest of their scientific achievements is in itself impressive, massive and invaluable. I am convinced that if Darwin or Wallace were alive today, they would wince in embarrassment at the dogmatism, the arrogance, the anti-scientific tactics and the bizarre reasoning of many modern day neo-"Darwinists." This should be a warning to all that the neo-Darwinists' assertions, much less their condescending arrogance, does not reflect upon the thinking and character of either Wallace or Darwin---not that they ever cite Wallace anyway. In fact, it is even embarrassing to refer to these gentlemen as "neo-Darwinists" and in so doing invoke Darwin's name in the process. A lot of the justifiable anger and scorn caused by "neo-Darwinists'" arrogance, rigidity and absurd diktats are misdirected at the inoffensive Darwin---for one thing by calling their stance "Darwinism." Charles Darwin was an eminently reasonable, intelligent, non-dogmatic, admirable scientist who, if alive today, would be embarrassed to the core of his being by many of the absurd claims made in his name and his theory by these gentlemen, not to mention their arrogant, condescending, obtuse, close minded, anti-scientific outlook.[20]

    Throughout his work, Darwin admits that there are holes in his theory due to a lack of data, which he hopes, and assumes, future scientists will fill in and verify his theory, but throughout, he openly admits that if not, then his theory is faulty. "If it could be demonstrated that any complex organ existed which could not possibly have been formed by numerous successive slight modifications, my theory would absolutely break down." (Darwin, 1859/1979; p. 146).

That is a true scientist.

Denton's Structuralism

Acknowledging Goldschmidt's contribution, Denton (2016) states that taxon-defining homologs (Types) occur suddenly, without any evidence whatsoever of gradual accumulation of parts in ancestral organisms, i.e., there are no intervening stages between species, no missing links, no empirical evidence; "the vast majority of all taxa are indeed defined by novelties without any antecedent in any presumed ancestral forms" (p.56). Denton points to hair, feathers, an enucleated red blood cell, a diaphragm, a laminar cerebral cortex consisting of six layers, the acquisition of language, the acquisition of mathematics, the endometrial stromal cell, the angiosperm flower, the tetrapod limb, the protein-coding genes and he details their complexity and their lack of antecedents. And, furthermore, that they occur in spite of the fact that they are often nonadaptive, that is, they do not confer any advantage to the recipient organism.

Denton then takes refuge in a structural alternative, wherein he states that there are "laws of biological form" very similar to physical laws and, in fact, the two go hand in hand. Certain biological forms will automatically occur at certain points due to intrinsic physical properties, somewhat like crystals in geology. As such, genes do not contain a blueprint per se, but the tools, the building blocks to achieve the necessary forms. This is a radical shift in paradigm, to put it mildly.[21]

But, then, where are these *Bauplans* (as he puts it) stored? Why these particular Bauplans? Why do some Types suddenly replace others? This structuralism does not really explain speciation either, i.e., how---and why---does one organism transmute into another. It is all very mysterious and unsatisfying.

Conclusion

Contrary to the neo-Darwinists' claim that all of the questions and criticisms of the classical theory have by now been answered and solved, it has by now, instead, become exceedingly obvious to all but the most intransigent that the classical theory of evolution is flawed, although the central concept of evolution remains intact. Although there are other shortcomings, the primary arguments, or rather evidence, that undermine the classical theory are: (1) speciation is not gradual, but sudden, saltationist, with periods of stasis in between; in other words, there is *no evidence* for the theoretical proposition that Natural Selection has resulted in speciation[22] (2) the taxonomic traits that differentiate one species from another are for the most part nonadaptationist, which begs the question as to how and why they arose and further undercuts the Natural Selection thesis (3) evolution has stopped; it is not a continuous process occurring in all species, as evidenced by the plethora of living fossils (4) many physiological components are so complex that their genesis cannot be due to a gradual accumulation of details, especially since there is no evidence for this proposition.

These arguments are unassailable.

It would be most appropriate to close this chapter with one quote from Charles Darwin himself

(1859/1979), a quote which is not often consulted by neo-Darwinists, but should be: "Furthermore, I am convinced that natural selection has been the main but not the exclusive means of modification." (p.68)

And while we are it, an ironic quote from Dawkins:

> There have been times in the history of science when the whole of orthodox science has been rightly thrown over because of a single awkward fact. It would be arrogant to assert that such overthrows will never happen again. (1986, p. 293)

FOOTNOTES TO CHAPTER EIGHT

The only bad theories are the ones that cannot be questioned or tested.
---James Lovelock

If the structural possibility is not there, you cannot have adaptation. Simply because out of nothing, nothing comes.
---Leon Croizat

[1]Carnivorous plants are invariably described as having developed their predation "because" of the poor soil in which they live, so that prey supplements their nutrient requirements, as if some animated, intelligent, plants had gotten together in a council and brainstormed a way to cope with this problem: "In some particularly impoverished environments---bogs, moorlands, mountain slopes washed by heavy rains---plants, to survive, have to feed on animals." (Attenborough, 1995; p. 72) Yet, numerous other plants grow and thrive in the very same soil, from trees to grass. I just recently returned from a collecting trip in Alabama and Mississippi. Along with the Miniature Huntsman's Horn, butterworts, and sundews, all in one spot, I also brought back over two dozen other plants and grasses that were growing on the same spot. None of them evolved a predatory mechanism. As an aside, carnivorous plants die if fertilizer is added to their soil or are planted in soil that is not acidic---which is definitely not adaptive.

[2]This is a new type of tropism, a carnotropism, in

addition to geotropism, hydrotropism, phototropism, unique to some of the active insectivorous plants.

[3]Although . . . we *know* that some diehard out there will *seriously* put forth the idea that some grazing animal somewhere, a sheep, a horse, a grasshopper, will become too scared to eat a leaf that has moved a little (say, by the wind) and will stampede away in a total panic. Such absurd assertions are common. I have come across books wherein the author, in all seriousness, asserts that, thanks to evolution and/or Natural Selection, the stripes on zebras serve as camouflage and break up their outline (Marler & Hamilton, 1966; Kruuk, 2003). Some authors state that the flat noses and slanted eyes of Orientals is an evolutionary product which helps them against the cold, even though the very same author, in the very same book, illustrates races with flat noses and/or slanted eyes that can be found in Australia, Africa, Borneo, Indochina and Philippines (Moore, 1964); what makes this even more absurd is the fact that Lysenko used this very same argument to bolster his pseudoscientific theories (Sudoplatov, 1994). And still another author who claims that the stripes of a tiger, viewed against the bright orange pelt, breaks up its outline and, thusly, serve as camouflage (Parker, 2003). Others have claimed that flamingos' and peacocks' coloration also serve as camouflage (Gould, 1991). Somewhere or other I have come across the statement that, when sea cucumbers are disturbed and they vomit their guts out, such behavior scares away predators though, to my knowledge, this has not been actually been empirically documented (and I, for one, have never witnessed it in the aquarium nor in the field). I am

certain that the reader, at some point, has come across a book or a documentary that has extolled the adaptive virtues of creating thousands, if not millions of progeny each season (e.g., seeds in plants, or eggs in invertebrates); likewise, I am certain that the reader at some other time has come across another book or another documentary extolling the adaptive virtues of creating just one, or two progeny, usually very large seeds, or avian eggs, with the rationalization that "energy" is focused on creating and/or nurturing just a handful of progeny. And who has not heard or read that the fact that some plants produce seeds so light that they are easily dispersed by the wind (while other plants' seeds are so heavy that they just drop down from the branches and stay in one spot---yet both types of plants thrive)? Or, the homily that poisonous animals, like the Gila Monster, the Monarch butterfly and the coral snake, are colorful in order to advertise and warn other animals that they are deadly and should not be bothered (while other animals like the cottonmouth, the black widow and the scorpion are not)?

As Zimmer (2001, p.147) has put it:

> Reconstructing the rise of these adaptations is treacherous work because sensible-sounding stories   about evolution are easy to make up. You see long tails on a swallow and decree that they must have evolved to let the bird maneuver more precisely, but someone else looks at them and decrees that they have evolved that way because female swallows find them attractive on male ones. Or maybe no adaptation is involved at all---maybe most of the swallows that happened to establish this species just happened to have long tails, and it's been that way ever since.

Along similar lines is the big brouhaha regarding altruistic vs. selfish behavior (Morell, 1996; Field & Brace, 2004). The logic behind the arguments for the supposed evolutionary basis for either style of behavior in animals and humans is pretzel-like. The reason is that both sides are wedded to the underlying assumption that the basis for either altruistic (or selfish) behavior is due to evolution, i.e., Natural Selection. Not so. Each style of behavior *just is*.

Biologists' obsession in looking for evolutionary adaptation at every single bit, every little trait, every little miniscule and global behavior and/or morphology invariably, predictably, results in some advantage being found out, even if it is tenuous, at best. The point is that if one is obsessed with finding something one will end up finding it, even if it is not actually there. The example that immediately comes to mind is the symbolism of the perverted crackpot Sigmund Freud; if you shared his obsessions, all sorts of interesting messages were revealed through his symbols. This is particularly the case with behavior; one can always attribute an adaptive advantage to an animal's behavior, just as one can attribute an adaptive advantage to the opposite behavior. For example, an animal that is omnivorous will have the advantage of dietary variety which maximizes the probability of finding food; animals that are food specialists (koalas, apple snail kites, panda bears) minimize interspecies competition and thus have an ample supply of food. No matter what it is, a neo-Darwinist can always attribute an adaptive advantage to anything, particularly if critical faculties are suspended.

This absurd attitude resulted in Continental scientists becoming hostile to the classical theory at the turn of the 20th century. This is why Continental scientists are not as enamored of the classical theory as the Anglo-Saxons (Reif, 1986).

(Yet, to date, as far as I know, no one has actually rationalized a selective advantage to an appendage that we all take for granted since it is ubiquitous: the tail. Almost all terrestrial vertebrates have a tail and, aside from birds and monkeys, the tails are almost always useless, being dragged around. This appendage has been handed down for eons as if it had an adaptive advantage, yet it has none, an excellent example of something that just *is*. It evolved long ago and it has been handed down from species to species to genera to genera to order to order.)

In all fairness, biologists are not the only ones guilty of compulsive rationalizing. I am thinking of the phenomena present in all cultures, from Indonesia to Brazil to Turkey to Cuba of women wanting to have lighter skin color and being lighter in color than the men they marry. This phenomena is present across ages as well, as we see in painted images of ancient India and the Roman Empire. The rationale was put forth by someone, whom I cannot now recall, that the desire for such a lighter skin color is because it represents a leisured life instead of working outside in the sun. No, it is just an inherent desire in human females.

[4]Mind you, to avoid the risk of beating a horse to death, we are going to omit the complex biochemical intricacies of, step by step, creating enzymes and nectar and the secretion of enzymes and nectar, which

processes, at any one step can be derailed if one step goes haywire.

⁵Does it? Although the process sounds very logical, to my knowledge there is no record of an individual organism within a species acquiring a new trait which results in the original species completely dying out. Besides, it has been many biologists' and ethologists' observations that in a population, marginal individuals do survive long enough to mate, not once, but several times. Discussing the rise of the dinosaurs, Benton (1984) makes a very good point: "There is simply no evidence that any large-scale competition ever took place, and even if there were, animals are too complex for us to say that one or another adaptation can by itself explain a major worldwide ecological replacement." (p. 56)

⁶Margaret Smith was her husband's right-hand man. She was once quoted as saying, "A wife can be independent or indispensable, but not both; I chose to be indispensable." (Thomson, 1991; p.28)

⁷Smith spoke the fantasy that many people of his generation had secretly nurtured when he said that he had always thought that somewhere, somehow, a survivor of the dinosaur era had survived, a fantasy that Sir Arthur Conan Doyle put into print with his *The Lost World*.

⁸It is no accident that the word "chauvinistic" is a French term. Around the turn of the century, when cave paintings were discovered in Altamira, Spain, French scientists denounced them as fraudulent. When similar cave paintings were discovered in French soil, they changed their tune once they realized that France had

had French cavemen. One can only wonder as to whether French cavemen were as obnoxious as their modern counterparts. Which brings to mind Dostoyevsky's famous observation: "It is a wonder how a people as inherently obnoxious as the French can nevertheless create beautiful art."

[9]Other living fossils that can be found today are journal editors, uniformitarians in the field of geology, the Nobel cabal members who decide who is *not* going to get a Nobel Prize, neo-Darwinists, and, of course, British Marxists.

[10]Another peculiarity of the Cambrian is that the organisms are already so complex. True, they are relatively small, with the exception of the meter long predator, *Anomalocaris*, but there is no denying that they are complex (the aptly named *Anomalocaris* was the *T. rex* of the Cambrian). Although *Marella* could be argued to be a predecessor of the contemporary trilobites, no fossils have yet been found of the earlier, simpler ancestors of many of the Cambrain fauna. Indeed, only sixteen species of the earlier Ediacaran fossils have been found. Of relevance, note that evidence has been found of large extraterrestrial impacts in the pre-Cambrian (Simonson, Byerly & Lowe, 2004) as well as some indirect evidence of the same during the Ediacaran epoch (Evans, et al., 2022).

[11] Let us ignore the fact that neo-Darwinians want to have their cake and eat it too: they simultaneously adhere to the above rebuttal while loudly proclaiming on the Internet and the journals and in books instances where evolution is alleged to have taken place much quicker than classical theory proposes (Weiner, 1994;

Chapman & Partridge, 1996; Roach, 2001; Hendry, 2000). Yet, animals like cats and ibises that were mummified by the Egyptians over four thousand years ago were found to be identical to their modern counterparts (Mayr, 2001).

[12]Stephen Oppenheimer (1998) has come up with an interesting theory (completely irrelevant to the present topic of evolution) wherein that there was a fledgling civilization in the Sunda Shelf when it experienced a sudden, cataclysmic flooding upon the ice sheets melting at the end of the Ice Age. A maritime people, its inhabitants scattered to the four winds, taking their legends, their customs, their food. Madagascar lies at the westernmost presence of the blowgun, an invention of the Sunda Shelf; it is also the westernmost presence of *Nepenthes* (and *Nepenthes* pitchers are looked upon as sources of medicinal properties by some rural Indonesians). In addition to the cultural aspects, Oppenheimer also points to the genetic marker links with Orang Asli aboriginals that have been found with Czechs and Kuwaitis. Two additional pieces of evidence that could possibly be used to buttress his theory are linguistic ones. As in the Indo-European similarities of names (Peter-Pete-Pedro-Pyotr-Piet-Pietro), there are many names in European and Middle Eastern ancient times that correspond to old-fashioned Sundanese names (e.g., Inna, Dudu, Mia, Musa, Amat, Kia, Enni, Otto, Herman, Nana, Ety, Didy, Lina, Risa, Dana). Secondly, just as the Bushman language has the unique linguistic "klicks," the Indonesian and Micronesian languages commonly use a unique reduplication (e.g., dada-chest, pipi-cheek, bibi-aunt, pingping-thigh, taktak-shoulder).

Well known examples are also aku-aku and Bora-bora. This linguistic peculiarity is also present in the Akkadian and Sumerian cultures of Mesopotamia (kaka-shout, qad-qad, head, gulgul-skull) (Heise, 2003).

[13]Orthodoxy states that the pitchers of the genus *Nepenthes, Cephalotus, Sarracenia, Darlingtonia* and *Heliamphora*---and each and every one of their respective species---with all the complexities involved in developing each step of their botanical predation, all owe their collective genesis to "convergent evolution" through Natural Selection.

[14]In fact, we see it today, to a lesser degree, in some of the great apes---which are herbivores---and in a highly exaggerated version with the tusks of the elephant. It may be argued that such elongation is adaptive because, for instance, elephants do happen to use their tusks. The simple answer is that if a mammal has a part of its component that it can use to its benefit, it will, whereas if it cannot do so, it will not.

[15]To my considerable amusement I subsequently found the very same, identical, argument in Lyell's (1833/1997) magnum opus, *except* that he was criticizing Lamarck's theory on the transmutation of species. In fact, I find it puzzling that Lyell should have been an ally of Darwin and Wallace at all, considering his very strong and lengthy opposition to Lamarck's principles of, first, the *progressive* system in organisms from the simple to the complex and, second, that "the descendants of common parents may deviate indefinitely from their original type." (p.184) Among other things, he makes an interesting point: if there is this progressive tendency for increased complexity in organisms, why is it that simple

organisms are still around?

At any rate, it is curious that over half of the second book of the *Principles of Geology* deals exclusively with the refutation of Lamarck's theory on the transmutation of species.

[16]And when humans intervene and hybrids are forced on the animals, the resulting hybrids tend to be sterile.

[17]Yet still another difference between the two totally alien organisms of plants and animals: the botanist Hooker pointed out to Darwin that variations in plants were exceedingly hard to quantify; variations, of course, are the *sine qua non* of Natural Selection (Weiner, 1994).

[18]A digression: I used to present to my students what I jokingly called Simon's Unproven, Unprovable, Law of Childhood Development. It used to go like this: "All babies are born geniuses but lose IQ points in direct relation to how dumb, or how neurotic, their parents are." It was mostly for the purpose of stimulating discussion, and some lively discussions did take place from time to time. Because it is un-provable, this "theory" is totally worthless.

Along the same line of pure speculation, in other words, a hypothesis, I would like to raise a point that, to my knowledge, has not been discussed at length: there is something fundamentally wrong with our view of early man. The emergence of *Homo sapiens*---depending on which authority one consults---has been placed at anywhere between half a million years ago to fifty thousand years ago. Yet, throughout that time, it was not until the end of the last Ice Age, with its accompanying extinctions, that humans began to form settlements of any sort from which sprung civilizations. There is wilderness all around and, suddenly, we have the Neolithic Revolution of 11,700 years ago when agriculture first developed, and advanced civilizations established around roughly 8,000 B.C. (7500-year old impact craters have been found in Estonia Raukas, 2022). There were settlements around seashores and riverbanks. Living near rivers was an obvious, logical choice. One never died of thirst even in times of drought and starvation was almost impossible: food was always available even if one did not know how to make a boat. More importantly, with time, one could see that seeds falling in the moist ground took root, hence the invention of agriculture. It was certainly a superior way of life than a nomadic one. And, once a settlement is established, civilization is not too far behind. And yet, for tens of thousands of years prior, humans do not seem to have formed settlements. True, we have unearthed areas where stone tool making took place, but they are not what we consider true settlements. And, also true that perhaps primitive settlements may have occur prior to 10,000 B.C. but they have not been found.

What about a possible alternative: what if the wiring in our brains was changed in some manner around the time of the end of the Ice Age---such as complex formation of language---that ultimately led humans to form settlements? Certain changes in morphology through evolution can be seen and touched in the fossil record, but changes in the soft tissues like skin color is another matter. Internal changes through evolution may occur in animals and humans and we would not be able to recognize it from the fossil record (after all, to take one example, let us remember that men and women, each as a group, differ in subtle and not so subtle ways mentally from each other because of how our respective brains are wired). Ten thousand years ago there was climactic stress and there was widespread extinction of the mammalian fauna (we are now at the very tail end of the Age of Mammals; mammals are becoming extinct nowadays at a rapid rate due to human activity; mammals truly flourished tens of thousands of years ago and the size of mammals then was impressive, whereas now the number of large terrestrial mammals can be counted in the fingers of one's hands). Of some relevance is the curious fact that not one word, not one, no matter how basic is the concept (food, mother, father, help) is common to all human languages.

Along these lines, I recently came across a paper which had a premise which, frankly, I found unnerving; it was like seeing for the first time a brick wall that had been in front of me all of my life. Ball (2017) points out the fact that our species has numeracy. Many animals can tell the difference in gross quantities of food, for example. Yet, humans can carry out division, multiplication, addition, square roots, quadratic equations, geometry, etc. We can, furthermore, grasp the abstract concept of numbers. And this ability has nothing to do with coping with the environment, with survival. If humans did not have this ability, the level of our civilization would be exceedingly primitive and stagnant.

At any rate, I have a very strong suspicion that the key that will decisively completely unlock the riddle of evolution lies in uncovering in what, exactly, took place "yesterday"---a mere 10,000 years ago---and not in the Cambrian, or in the K-T boundary. Approximately 10,000 years ago there was a sudden turnover of species, when gigantic floods took place, when other hominids vanished, the Ice Age ended, when man probably changed physically into races and, most, importantly, got down to the business of building civilizations after tens of thousands of years living like animals; it seems to be that if we are going to crack the code of extinction and evolution, our best chances are in investigating what happened "yesterday").

And---again, this is pure speculation, or to put it scientifically, a hypothesis---is it possible that a small comet(s) crashed into earth at that time, whose impact caused a terrific heat blast(s) that melted the ice, resulting in a worldwide flood? It might explain the ancient European superstition of (beautiful!) comets being harbingers of catastrophe. Pure speculation, of course. Except that evidence has been found (Kennett, et al., 2009; Israde-Alcáantara, et al., 2012; Wittke, et al., 2013) of multiple extraterrestrial impacts during the Younger Dryas by a fractured comet (most comets lack iridium, incidentally). These impact(s) have been disputed by others (Pinter & Ishman, 2008).

Furthermore, extinctions during The Age of Mammals have been absurdly attributed to overhunting by humans, particularly in the New World. While it could apply to a few individual species, it is unrealistic to attribute it to humans. Herbivores in islands may lack predators and become easy prey for introduced predators, as was the case with the dodos and Taínos, but this does not apply to continental herbivores who had to contend with powerful predators prior to the arrival of humans. Mammoths, mastodons, glyptodonts, giant sloths and giant camels had to contend with saber tooth tigers and giant hyenas---and there is nothing more vicious than a hyena, regardless of size. Besides what killed the giant predators? Therefore, the arrival of humans and the extinction of large mammals must be seen as a correlation, or a parallel event and not as a cause and effect relationship. Besides, if the theory of man as a superpredator is true, why do we still have bison and bears? Why are elephants and gorillas and rhinos still around---now that Africans have rifles? Even now, with our superb weapons and our vehicles, they are still hanging on---granted that they are on the verge of extinction, thanks to the avaricious Chinese, the point is that for the past century and a half of the introduction of superb weapons and vehicles, they are still here. The theory that it was Man that was the one that wiped out the giant mammals is as absurd as the old theory that it was the egg eating rodents that killed off the dinosaurs---even the oceanic ones.

[19]For example, from the female viewpoint: "And one thing I certainly didn't grasp was that the girls in these penniless middle-class families will marry anything in trousers, just to get away from home." (Orwell, 1939; p. 156)

[20]If this sounds too harsh consider just one instance: *New Scientist* published a cover story "Darwin Was Wrong," which argued that the traditional phylogenetic tree should instead be a web; neo-Darwinists, including Richard Dawkins, the foremost

contemporary neo-Darwinist, called for a permanent boycott of the magazine.

[21] The question of novelty, that is, the rise of brand new, unprecedented traits has, since the 1990s, become the focus of evolutionary biologists since it was practically ignored in the Modern Synthesis, which focused on adaptation, variation and population dynamics, and the use of statistics. It was obvious, after all, that Natural Selection could not influence traits which had not appeared (Müller, 2010; Müller & Newman, 2005).

[22] Richard Dawkins published numerous books devoid of any replicable, incontrovertible, experimental evidence that support the idea that Natural Selection is the mechanism behind evolution. At best, he has conjured up squiggles made by his computer, which he points to as proof (Dawkins, 1986). Indeed, as often stated above, one of the characteristics of this field are the flights of imagination in conjuring up the past, present and future instances, and circumstances, of evolving species that have absolutely no actual basis in reality, but which are touted as factual. *The Extended Phenotype* is just one such instance, a work that Dawkins refers to as his best work yet (Dawkins, 1982). The present author agrees with the assessment.

CHAPTER NINE

THE MCCLINTOCK EFFECT: THE SEVENTH MECHANISM

Stasis is data.
---S.J. Gould & N. Eldredge

Thus extinction and natural selection will, as we have seen, go hand in hand.
---Charles Darwin, *The Origin of the Species*

When you have eliminated the impossible, whatever remains, however improbable, must be the truth.
---Sherlock Holmes

"All truths are easy to understand once they are discovered; the point is to discover them."
---Galileo Galilei

Cataclysms

Imagine that you are a Mayfly. Your lifespan as an adult is one day, no more. If you emerged on a rainy day, you would think that it always rained on this world. If you emerged on a hot, sunny day, you would think that this world was always sunny and hot. If on a very windy day, the world would be perceived as always blusterous and if on a snowy day, you would think that it always snowed.

As it is with Mayflies, so it is with humans. Continuing with the metaphor, some people are of the firm opinion that since nothing extraordinary has happened in their personal lifetime then nothing extraordinary *can* happen. They have not been personally involved in avalanches, tornadoes, hurricanes, volcanic eruptions, or war; consequently, those events cannot occur. As Chapman (2002) succinctly put it, "What are very tiny risks for impacts during a human lifetime become certainties on geologic time scales."

We are living in an interlude between Ice Ages. During the Pleistocene alone, there have been approximately half a dozen Ice Ages. For the most part, in our present interlude, the climate has been very pleasant. Nor have we been visited by cometary bombardment or massive geological upheavals. True, Tambora erupted in 1815 (which eliminated summer for a year), Krakatau in 1883 (Winchester, 2003), Mt. Pelee in 1902, Mt. St. Helens in 1980, Mt. Pinatubo in 1991 and Mt. Vesuvius in 79 A. D. (Pliny, 113/1969), but these were all mere firecrackers, nothing at all like the previous eruption of Tambora or of Krakatau that severed Java from Sumatra, thereby creating two islands out of one, or the gargantuan caldera in the same archipelago. And, true, some astronomical object or other exploded in the atmosphere in Siberia, somewhere over Tunguska in 1908, but, it, too was a firecracker and, most importantly, it occurred over an uninhabited area, so, as mankind was concerned, it did not really happen; if a meteor falls in the forest and there is no one to hear it, has it really fallen, has it really made a sound? (as an aside, I distinctly remember when the Tunguska event was dismissed as being the equivalent of astrology or the Bermuda Triangle).

S. J. Gould (1982) wondered why, exactly, had gradualism taken such a tenacious hold on paleontology (with the parallel phenomena in geology of uniformitarianism) when there was absolutely no data to correspond with the degree of stubbornness encountered and concluded that it must be due to a psychological, or cultural, factor. This would not have surprised Thomas Khun one bit (Hull, 1996). Note the *zeitgeist* from a geologist (Benton, 2003, p.71):

> A key feature of the debate about mass extinctions, from 1840 to 1980, has been the imbalance in perceptions of the two sides. The proponents of catastrophe and sudden mass extinction were consistently regarded as lunatics. To link a mass extinction to cosmic rays, sunspots or meteorite impacts was to class yourself with the pseudoscientists and astrologers. The extinction-deniers were the level-headed, careful scientists.

There is no question that it is a matter of temperament, of acculturation. The explaining away of the demise of the dinosaurs by the gradualists is particularly amusing. As a boy, I remember that the disappearance of dinosaurs was attributed to various causes. One of them hypothesized that small mammals might have eaten the dinosaur eggs which the adults would have been unable to prevent because they were too stupid. Even to my young mind this made no sense at all if for no other reason that it did not explain why the aquatic dinosaurs were no more (I also remember that there was a rival theory, put forth by one de Laubenfels, prior to the discovery by the Alvarez team, that perhaps an asteroid, or comet, may have been at work). Even more absurd theories have been put forth to explain away the demise of the dinosaur, from sudden infrared heating (Archibald, 2005) to one theory put forth by a Frenchman blaming accumulated dinosaur flatulence (Benton, 2003). (And such was the influence of this gradualist mental set that the occurrence of a fossil was described as an animal dying, whereupon, very, very, very gradually, it would be covered by sediment and the organic materials replaced by minerals. Whomever put forth this scenario must have lived their entire life in a big city. In Nature, any carcass is immediately pounced upon and devoured and mutilated by scavengers, great and small. In reality, most fossils were instantly buried at the moment of death, almost certainly by a local cataclysm; a relatively recent find of a *Velociraptor* battling a *Protoceratops* attests to this).

Then, at the end of the 1970s, the Alvarez team found a layer of iridium at the K-T boundary and since iridium is an isotope found in extraterrestrial objects, factual evidence was found to support an impact hypothesis (by the way, Alvarez's (1997) account reads like a fascinating piece of detective work---which it was, of course). Gradualists were very upset. The subsequent finding of iridium and fullerenes in deposits at the same geological boundary in New Zealand, Denmark, India, Bulgaria and other locations, as well as the finding of tektites and of tsunami deposits led the team to the smoking gun, or rather, bullet hole, the Chicxulub Crater in the Yucatan peninsula, as well as circumstantial data for the impact (Alexander, 1981; Rensberger, 1986; Bohor & Seitz, 1990; Lessem, 1992; Alvarez, Clatys & Kieffer, 1995; Hildebrand, Pilkington, Connors, Ortiz-

Aleman & Chavez, 1995; Schuraytz, et al., 1996; Roach, 2002; Parthasarathy, et al., 2002; Preisinger, et al., 2002). It is now known that the object that slammed into the Earth was bigger than Mt. Everest, was traveling at a tremendous velocity (Alvarez, 1997) and was a meteor, not a comet (Kyte, 1998). Gerta Keller of Princeton has also put forth the hypothesis, that it was not just one impact, but multiple impacts, similar to what happened to Jupiter with Comet Shoemaker-Levy 9, a hypothesis supported by evidence (Keller, el al., 2002).

At any rate, the impact caused massive shock waves, earthquakes and tsunamis (Matsui, et al., 2002; Norris & Firth, 2002). Fiery ejecta caused forest fires (Kring & Durda, 2003), the massive soot being recorded in the geological strata and estimated at 10% of the world's biomass having been incinerated. Fiery fragments of the meteor fell thousands of miles away. What made things particularly bad for the terrestrial dinosaurs was how little land was available at the time, compared with now; the oceans had covered many continents. About a third of North America, South America, Africa and Australia for example, were under water (Bakker, 1986). A dust cloud affected the atmosphere globally, exactly to what extent is unknown (it must be remembered that the mild volcanic eruptions during our lifetimes have affected visibility). But it must have been exacerbated by the smoke from all the massive forest fires, although it has also been postulated that the gases chlorine and bromine were created, which damaged the ozone layer. As if that was not enough, the impact area was rich in sulfur, which resulted in acid rain (the equivalent of lemon juice (Kerr, 2013)), making a bad situation much worse (Braun, 2002). A cascading effect would have taken place, obliterating whole ecologies (Artemieva, 2017), although there is some tentative evidence (Bakker, 1986; Stanley, 1987; Fassett, Zielinski & Budahn, 2002) that suggests that some dinosaurs did not all die out right away, but that there were a handful of terrestrial species that lingered for a few millennia in microclimates before experiencing a population crash (the role of checkerboard microclimates (Stuller, 1995) in evolution is important). Consequently, it has been argued that the impact was not an extinction event. However, what is forgotten in this argument is that if, indeed, some species of dinosaurs survived for a while thereafter, they were simply *living fossils* that ultimately died out, much like the modern-day coelacanth, which has not died out.

Even so, the inescapable facts remain: first, not one dinosaur---of any size---in the end survived. Second, all huge animals died out. Third, even foraminifera and phytoplankton became extinct (Gallagher, 2002; Huber, MacLeod & Norris, 2002). Fourth, several ectothermic scavengers survived---it was easy pickings. Fifth, the hitherto suppressed mammals flourished.

Although Alvarez and others credit the impact with the final extinction of the dinosaurs, gradualist paleontologists are loath to admit it (Bakker, 1986; Stanley, 1987). Bakker, in particular, has proposed a truly absurd theory to counter this, which is all the more surprising when one considers the innovations that he has promoted, such as that birds are the descendants of the dinosaurs---which, in actuality, T. H. Huxley first put forth in the nineteenth century (Jaffe (2000)---although at the same time, it must be admitted that Bakker is also the one who is responsible for the recent portraying of bipedal dinosaurs with their tail up in the sky and their center of gravity being so forward that their snouts are almost scraping the ground. The reason that the dinosaurs became extinct, according to him, is that the climate changed and land bridges formed between continents, which allowed nonnative disease-carrying animals to intermingle. But, in every epidemic there are always at least a few survivors, yet no dinosaur at all survived, not even on islands. And what about the marine fauna? No. Additionally, this relates to the so-called Great American Interchange of much later during The Age of Mammals, that is, that the connection of the American continents via Panamá is cited as the basis for the mammalian mass extinction: it is assumed that those animals were at the border just waiting for an opportunity to cross, a migration similar to the Sooners in the Oklahoma land rush. Yet, how many jaguars do we find in Wyoming? How many grizzly bears do we find in Patagonia? No.

Some have put forth the proposition that dinosaurs were well on their way to extinction when the impact accelerated the process. They present qualitative arguments for sea level change (Raup, 1992). Yet, every *quantitative* stratigraphic field study has concluded that biodiversity was unchanged up to the time of impact (Fastovsky, 2005).

Aside from the impact itself, there was a simultaneous vast output of lava in India, called the Deccan Traps. This has resulted in another absurd either/or argument as to which one was really the one that killed off the dinosaurs, the impact, or the lava flow, astronomy vs. geology, since gargantuan volcanic outpourings have occurred at the same time as mass extinctions (Kerr, 2000). But this is yet another instance in science of false dichotomies. Obviously, it was both. There are three possibilities to this scenario. First, either there was a multiple impact, one of which landed on India with such particularly excessive power that it shattered the crust for a long time, whereupon lava poured out for years and covered up the crater, in which case Chicxulub was a minor fragment. Or, the Yucatan impact arrived at a different time. Or, equally possible, the Chicxulub impact caused the Deccan Trap, which was antipodal to the impact zone. Regardless, keep in mind that a bullet, after all, is simply a pebble that weighs just a few grams and that asteroids range in size from a speck of dust to a minor moon; the compressed air of a descending meteor increases the impact's power. Those advocating volcanism as the only cause of the dinosaurs' demise (Kerr, 2000) have never explained in a truly convincing manner (Czamanske & Fedorenko, 2017; Self, et al., 2022) what caused the basaltic traps in the first place and why at that particular time, disrupting the geological homeostasis.

Flood basaltic eruptions just happen.

Like spontaneous generation.

Regardless, the bolide, along with the Deccan traps that resulted in a nearly continuous outpourings of volcanism in a span of around 2 million years, released toxic chemicals and gases (Gerasimov, 2002; Skála, et al., 2002; Keller, et al., 2020; Eddy, et al., 2020; Kale & Pande, 2022; Cox, & Keller, 2023) into the atmosphere which must have affected organisms. On top of that, freshwater ecosystems experienced anoxia Gardner & Gilmour, 2002).

(I mentioned the mayfly perspective earlier: we have the delusion of being unaware that the biosphere actually rests on an eggshell. The Earth's equatorial diameter is 12,756 km/7,926 mi. Russian scientists at the Kola peninsula attempted to drill through the earth's crust: *the earth's crust is a mere 25 miles deep (40 km)*. That is less than half the distance from Austin to San Antonio, less than a half hour's drive by car! Further, take also into consideration that in many parts of the planet, the Earth's crust is somewhat hollow; by this I mean that there are gigantic, cavernous, aquifers. They are geological bubbles. The Ogallala Aquifer covers eight states in North America, while in South America the Guaraní Aquifer includes Uruguay, Paraguay, Brazil and Argentina with a depth of 1800 meters.

A large enough meteor travelling at the speed of a bullet and at the right angle can shatter and penetrate the earth's crust at the point of impact. Others cannot, depending on the size of the asteroid, the speed, and the angle of trajectory: for example, the Chesapeake Bay impact crater was only able to penetrate to 5 mi/8km of the surface.

The point to keep in mind is that the earth is a gigantic ball of churning, molten lava, superimposed by an eggshell-thin cooled crust on which life lives. It helps to envision this if one thinks of a ball whose diameter is the size of a ten-story building covered by an eggshell. If this is difficult to grasp, keep in mind that in many parts of the world (e.g., Germany, Japan, America) where there are hot springs which are resort spas).

The Permian Extinction

Something much worse occurred during the earlier Permian extinction (Benton, 2003).

Depending on whose authority one consults, 90-96% of the species living at that time died, when the Earth became a Dustbowl. The increase in temperature corresponded with hypoxia in ocean waters Penn, et al, 2018). The Permian mass extinction was the worst that ever occurred, at which time, life came within a hair's breadth of vanishing from Earth (Erwin, 1996). Simply put, the planet became a hellhole.[1] The Permian Siberian Traps are similar to the Deccan Traps, only much bigger (1.5 million cubic kilometers, truly gargantuan, about half the size of China). Of relevance is also the presence of isotopic anomalies, including strontium (Renne, *et. al.*, 1995). High concentrations of mercury and carbon isotope have also been found (Wignall, et al., 2009; Font, et al., 2016), as well as SO_2 and CO_2. A crater, named the Bedout High, was found in Australian waters that has become the primary candidate (Becker, et al., 2001; Becker, Poreda, Basu, Pope, Harrison, Nicholson & Lasky, 2004; Wright, 2005). A more recent discovery of a three-hundred-mile crater in the Wilkes Land region of the Antarctic, under tons of ice, (Weihaupt, 1976; von Frese, et al., 2009)---three times the size of the Chicxulub crater---is another rival for the Permian extinction, along with the Siberian Traps. It is possible that a meteor or comet did not impact in the site of the Siberian Traps; von Frese, et al., (2009) suggest rather that the Traps were caused by the Antarctic impact since the site of the Traps was antipodal, not an unusual hypothesis if one keeps in mind that a bullet will make a deeper exit wound than an entrance wound. They also point out that volcanism have occurred antipodal to impacts both in the moon and in Mars. On the other hand, Koeberl (2007) Köberl, et al., (2004) have disputed the evidence for impacts for the Permian extinction based on geochemical evidence.

Other Impacts, Other Traps

It must also be kept in mind that although the Permian extinction and the dinosaur extinction, along with their Siberian and Deccan traps, get the most press, numerous mass extinctions have taken place as well. And, concomitant basaltic eruptions have been present, on an average one in every 20 million years, according to Courtillot & Olson, (2007) (e.g., Columbian River and Chilcotin Plateau Basalts (late Miocene/early Pliocene), Caribbean Large Igneous Province, Emeishan Traps, Skagerrak-Centered Large Igneous Province, Ethiopia-Yemen Continental Flood Basalts, North Atlantic Igneous Province (late Paleocene/early Eocene), Ontong Java Plateau, Rajmahal Traps, Paraná-Etendeka Traps, Bunbury Basalt, High Arctic Large Igneous Province, Karoo, Franklin Large Igneous Province, Ungava Magmatic Event, Mackenzie Large Igneous Province, Viluy traps, Kalkarindji Province, Bonarelli Event, Faroe Islands Basalt Group, Central Atlantic Magmatic Province. Then, there are the Libyan Desert glass field and the Dakhleh Oasis glass field in Egypt. In regards to the North Atlantic Igneous Province, which occurred at the Paleocene-Eocene boundary, Wignall (2015) makes the following observation: "On the whole, this was a time of evolutionary success for a broad range of animals. Mammals did especially well. Many modern families appeared, including our own group, the primates, together with the first members of the deer and horse families." (p.162) He (Wignall, 2001) is one of several who have pointed out the correspondence of extinctions with flood basalts as well as occasional impacts.

Gradualists, though, are of the firm belief that the only time that the Earth was subjected to impacts was at the beginning of its existence. And, as absurd as it may sound, Meteor Crater in Arizona was once pronounced by the Chief Geologist of the U.S. Geological Survey to have been produced by volcanism (Benton, 2003); Gene Shoemaker finally laid that ghost to rest, then proceeded to prove that the town of Nordlingen, in Bavaria, does, indeed, lie within a crater (Shoemaker & Chao, 1961).

For a long time, they even *refused* to believe that the planet's magnetic field had reversed itself when Matuyama proposed it.[2] But that is another story.

The Chicxulub Crater impact aside, it was not the first, nor the last time, that Earth sustained a collision with an astronomical object. Although uniformitarians readily admit that there was meteoric bombardment during the planet's very beginning, they refuse to even consider that any major impacts have occurred since then. They have no evidence, they just do not want to believe it; catastrophists, on the other hand, do present data; for example, the Eocene-Pleistocene transition is identified with several impact craters and a limited biological extinction, which corroborated some astronomers' computer model that showed a star passing through the Oort Cloud, altering the comets' orbits therein (Weissman, 1988). As for any future impacts, they state with conviction that the chances are astronomical that it may happen, so it will not.[3] It is apparent that nothing will change their minds unless a meteor falls down on their heads (actually, it is known that star Gliese 710 will be plowing through the Oort Cloud in the distant future (Wessman, 1988)).

The world's *zeitgeist* changed radically with, first, the Alvarez *et. al.,* finding and confirmation and, second, the dramatic Shoemaker-Levy comet. Many finally begun to realize that the planet has indeed sustained numerous, powerful collisions throughout its history; those individual researchers, instead of debating with the gradualists, have simply bypassed them and have been looking for evidence of additional impacts on Earth; the data has begun to steadily drift in in support of this view; in 1964 only five or six terrestrial impact craters were admitted being in existence. Once the blinders came off over one hundred have been identified (e. g., Weissman, 1988; Kerr, 1995(a); Köberl, Poag, Reimold & Brandt, 1996; Raup, 1999; Harder, 2002; Williams, 2004). Predictably, this has resulted in the pendulum swinging the opposite way (Reimold, 2007), even to the point of meteor/comet impacts being part of pop culture.

For instance, in Quebec, we have the enormous Manicouagan crater, 700 kilometers in diameter, which is presently, curiously, an annular lake (see Figure). It is believed to have been formed 200-214 million years ago; the Permian mass extinction took place 250 million years ago (actually, in reality there is great uncertainty as to the exact age of the boundary (Rene, 1995)), otherwise, it would have been a perfect candidate as the culprit of the extinction. Manicouagan is the fifth of a multiple impact meteor, three of which are in the same latitude (Rochechouart, France; Manicouagan and Saint Martin, Canada); the other two fell in Red Wing, United States and Obolon, Ukraine (VanDecar, 1998; Spray, Kelley & Rowley, 1998), not too distant in latitude from the first three; the Siberian Traps are also on the same latitude.

Impact(s) could just also be the candidates for the mass extinction at the end of the Triassic. The dinosaurs' rise and dominance are linked to an asteroid hit; in a mere 10,000 years, dinosaurs abruptly grew by 40% and more than doubled in mass while at the same time period the rauisuchians disappeared; just as telling, we find that there was also present an abundance of fern spores and iridium (Olsen, et al., 2002; Winters, 2003). Marine and continental extinctions were simultaneous (Whiteside, et al., 2010), as were the extinctions for tetrapods and invertebrates (Olsen, Shubin & Anders, 1987). In other words, the dinosaurs were ushered in as the dominant organism because of an astronomically based mass extinction and they exited because of another astronomically based mass extinction (Parsell, 2001). The T-J extinction boundary is synchronous with the gargantuan Central Atlantic magmatic province, the biggest basaltic flow (Olsen, 1999), and the Manicouagan impact in Canada (although Hodych & Dunning (1992)) disagree on Manicouagan's role. "The rapidity and synchronism of the Triassic-Jurassic continental and marine extinctions is difficult to explain by simple gradual mechanisms." (Olsen, Shubin & Anders, 1987; p. 1027)

Benton (1984) and Frederick & Gallup (2017) have put forth the notion that it was not an impact but a change in vegetation, which served as the basis of nourishment for the rhynchosaurs, that led to the rise of the dinosaurs, but, again . . . what about the marine life? Benton does make one good point which is relevant to our overall discussion in this book: "There is simply *no evidence* that any large-scale competition ever took place, and even if there were, animals are too complex for us to say that one or another adaptation can by itself explain a major worldwide ecological replacement" (italics mine; p. 56).

Apropos, I will digress here yet again: listening to neo-Darwinists (and others, in all fairness, such as Knoll (2003)) talking or writing about mass extinctions, they describe ecosystems then being available to the surviving species which *magically* evolve into new species. No rationale whatsoever, no data in any matter whatsoever, is given as to *why* radically new species should magically appear instead of the old surviving ones simply occupying the now vacant niches. They simply state that . . . it just happens. Presently, in the Caribbean, there are occasional attempts by island nations to build artificial reefs by sinking large inorganic structures, from derelict ships to blocks of concrete and granite. These objects are very quickly colonized by fish and invertebrates which desperately seek a place of refuge---since most of the ocean is an aquatic desert just as the poles are an icy desert and the Sahara is a sandy desert---from sponges and sea quirts to starfish, eels, octopi, crabs and fish. Notice that in each instance of the numerous artificial reefs not a single new species has emerged in any phylum. Now, contrast this fact with neo-Darwinists' assertions that, after a major extinction of the faunal and flora, the surviving species, in filling in the newly vacated niches, decide to undergo radical physiological alterations into new species, that the newly vacated ecology simply results in evolution taking place. Such assertions are devoid of *any* supporting data, much less experiments, but are often made whenever the topic of mass extinctions crops up. Now, likewise in the Caribbean (to beat a sea-horse to death), the poisonous red tide occasionally decimates entire ecosystems, leaving them open to opportunistic squatter organisms (Grant, 2019). At no point has speciation ever occurred as a result of such a vacant ecosystem, as some claim happened after, say, the Jurassic mass extinction. And the same is true on land, of course. Similarly, during the 1930s, North America experienced a crippling drought, resulting in the Dustbowl, yet no speciation occurred. Africa at present has been experiencing decades-long drought and, again, there is no speciation in sight. Another example: the sites in Holland and Great Britain where Niko Tinberger was carrying out his ethological studies on seagulls over several decades experienced changes in its ecology, from invasion by predators, specifically foxes, and an increase in floral diversity; no speciation took place. And, lastly, when Krakatoa decimated all life in the remaining islands, the colonization that began to take place was by opportunistic species. Again, no speciation took place.

In the case of human beings, when a high-rise condominium suddenly becomes vacant, people do not suddenly start sprouting wings in order to get to the top floors, they just move in.

But to return to the immediate subject at hand:

Canada alone has six impressive impact structures, not counting the offshore Montagnais crater near Nova Scotia (Erickson, 1991).[4] Just as dramatically, if not more so, is the medieval town of Nordlingen, in Germany, mentioned above, which is located inside a meteor crater; its church was partly built with blocks of impact breccia (Conway Morris, 1998; Benton, 2003). There are at least *one hundred* impact craters that have been identified on Earth; almost certainly there have been many others which are no longer in evidence due to erosion and subduction. Unfortunately, one of the problems is that impact craters are notoriously difficult to date with exactitude (Raup, 1999). However, no crater has, as yet, been found and definitely credited for the Permian extinction and may never will be for the simple reason that the Earth is 70% covered by water and the crater (or craters; there could have been half a dozen impacts at the Permian extinction) may have been obliterated when oceanic crust is subducted into the mantle and/or through erosion. Nonetheless, an extraterrestrial impact is the only logical answer. Earth does not exist in a vacuum.

Of additional relevance to this discussion is the moon. Earlier in its history, the moon was much, much closer to the Earth and has been steadily, gradually, expanding its orbit. During its earlier history it almost certainly served as a shield from bombardment because of its proximity, hence its hundreds of craters (such is the influence of uniformitarians that for some time, it was thought that the craters were volcanic rather than from impacts (Raup, 1999). Nonetheless, the Earth did sustain heavy meteorite bombardment early in its history (Kamber, 2002). The far side of the moon is featureless in that it is uniformly covered with countless craters (similar to the other saturation-cratered moons in the system and to Mercury (Guterl, 2004)),[5] whereas the side facing the Earth has several gargantuan *maria* (read: basaltic traps). There is at least one recorded observation of a meteorite impact on the moon during historical times (Erickson, 1991).[6] Daichi Fujii, curator of the Hiratsuka City Museum, recorded another impact in 2023.

Even more importantly is the fact that iridium anomalies have been found at the Eocene-Oligocene boundary, the Permian-Triassic boundary, the Cambrian-Pre-Cambrian boundary, the Devonian-Carboniferous boundary (which confirmed the Canadian paleontologist Digby McLaren's ignored 1970 prediction of a meteorite impact causing the Frasnian extinction) and towards the end of the Jurassic---the Brochwicz-Lewinski "event," wherein another mass extinction took place. Microtektites have also been found in some of the boundaries whereas in others there have been magnetic spherules similar to those found in Tunguska (Raup, 1999; Perkins, 2008). That impacts occurred during the glacial period with catastrophic results for the climate and the fauna is a certainty (Johnsen, et al., 1997; Zamora, 2017; Kjaer, 2018).

As to the possibility that our planet would again sustain in the future another impact by a meteor, or a comet, the question was simply waved away. Occasional, repeated close calls by meteors almost hitting us have made absolutely no impression on them. For example, literally as I was writing this passage, I just now became aware that a close encounter took place three months ago; no one noticed it until it was three days away, a mere 26,500 miles away (Tyson, 2004).[7] One should keep in mind Smith's Law for comets: "Anything that did happen, can happen." In 1970, Singer (1970) proposed that Venus' retrograde motion was due to the planet's collision with a moon, an unusually speculative proposition, while Nimmo, et al., (2008) suggested that the reason for Mars having one hemisphere thinner is due to massive impacts in the past.

And then, in 1994, reality set in with a vengeance as the Comet Shoemaker-Levy 9 slammed into Jupiter (conveniently on the 25th anniversary of the Apollo landing on the moon) in a series of impacts that, luckily, was dramatically recorded by instruments. Fragment G alone, which was average, created a ring of hot gas 33,000 kilometers wide, wide enough that if it had fallen to Earth, *it would have encompassed the northern hemisphere*; it expanded at four kilometers per second from the center of the G spot (Levy 1995) and created a fireball that was sent thousands of miles into space and lasted for well over half an hour; if it had occurred on Earth, it might just have depleted the oxygen level. The fragment was, at most, only two kilometers in diameter. For those people in denial about the probability of our having been impacted in the past, or, about the possibility in the future, they must have experienced cognitive dissonance something awful; however, many other person's opinions at least about the K-T bolide, were changed, since the meteor that wiped out the dinosaurs was much larger (Melosh, 1995).

The major mass extinctions of species on Earth have been caused by extraterrestrial impacts which in turn, caused massive volcanism/basaltic. Impact craters may not always be found corresponding to the boundaries (in fact, to a certain extent, impact craters are irrelevant in so far as they did not shatter the crust, resulting in flood basaltic eruptions which covered up the crater). However, it must be kept in mind that the Earth is 70% covered by water and the crater may have been obliterated when oceanic crust is subducted into the mantle and/or through erosion, or, much more likely, buried under basaltic flow if the crust is compromised. Nonetheless, an extraterrestrial impact is the only logical answer. Although iridium is usually absent in comets and not always present in meteors, iridium anomalies have been found in six of the geological/paleontological boundaries; the demise of tropical reefs have also been found at those boundaries and invariably follow mass extinctions as has flood volcanism and spikes in fern pollen (Raup, 1992). Incidentally, it may be remembered that when Krakatoa exploded ferns were the first to colonize the devastated island (Quammen, 1996; Winchester, 2003).

It needs also to be pointed out that the Comet Shoemaker-Levy 9 was the only object impacting another planet that was ever detected from Earth.[8]

It is interesting that as early as 1767, it was

recognized that a comet coming to Earth might result in unpleasant consequences:

> When the movement of the comets is considered and we reflect on the laws of gravity, it will be readily perceived that their approach to the Earth might there cause the most woeful events, bring back the universal deluge, or make it perish in a deluge of fire, shatter it into small dust, or at least turn it from its orbit, drive away its Moon, or, still worse, the Earth itself outside the orbit of Saturn, and inflict upon us a winter several centuries long, which neither men nor animals would be able to bear. The tails even of comets would not be unimportant phenomena, if the comets in taking their departure left them in whole or in part in our atmosphere.
> ---Lambert, *Lettres cosmologiques* (Sagan, 1977)

There have been other cataclysms besides asteroids and comets. The most substantiated one was the gigantic flood that took place at the dawn of human civilization, approximately 10,000 years ago, and which affected the entire world, when the Ice Age abruptly ended. The event has survived and been passed down in human tales of religion and poetry. It was truly a gargantuan, sudden, permanent flood. It flooded the subcontinent of the Sunda Shelf, for example, turning the area into the present Indonesian archipelago. In the Persian Gulf, a gigantic wall of water rushed all the way up to what is now past Basra. Elsewhere, it affected the local geology (Bretz, 1969; Gould, 1982; Oppenheimer, 1998). Regionally, it has been known by various names, such as the Spokane Flood (Albee, 2004). The sudden jump in sea level due to meltwater has been objectively verified and dated from the fossil record and, in actuality, there were two major floods, one 13,800 years ago and another 11,300 years ago, and a minor one 7,600 years ago with no reversals (Bard, *et. al.,* 1996).

Three more, random, isolated cataclysmic events in early human history:

In Asia, Sumatra and Java were once joined, but an explosion from Krakatau, which dwarfed the one in 1883, severed both parts and the ocean rushed in. This occurred during human times and the event was recorded.

In Europe, the Black Sea was a huge freshwater lake separated from the salty Mediterranean by a sliver of land. An earthquake, one of endless earthquakes in that part of the world, collapsed the barrier and a sudden, gigantic flood rushed into the lower lever and wiped out the human inhabitants.

In, roughly, 1628 B. C., the inhabited island of Thera (or Santorini) began a series of eruptions which ultimately ended in a cataclysmic explosion which vaporized two thirds of the island (there is a caldera, in the sea floor right next to the island). It created an enormous deadly cloud which spread southwards. It also created tsunamis which at least one person has put forth as the cause of the destruction of the Minoan civilization. It has also been suggested that the tsunami may have been recorded in Exodus; when the Jews fled Egypt, the seas parted, only to come crashing down over Pharaoh's army: whenever a tsunami occurs, the water near the shore quickly is sucked out towards the deeper end just before the tidal wave arrives (Sagan, 1977; Pellegrino, 1991; Refrew, 1996).

I hope that by now the reader will acknowledge

that, occasionally, cataclysms have occurred.

Ice Ages

To a certain extent the widespread glaciations known as ice ages can be considered to be catastrophic in so far as whole ecologies are devastated upon the incipient glaciation. Nonetheless, glaciation supposedly take place incrementally. The earliest global catastrophe to affect living things (Mayell, 2001) occurred when almost the entire world became glaciated, the Snowball Earth theory (Donnadieu *et. al.,* 2004; Kerr, 2005). When the global glaciation ended, the proponents of the Snowball Earth theory point out that the dearth of organisms prior to the Cambrian and the proliferation of (complex) organisms in the Cambrian (the so-called Cambrian Explosion) paralleled with global glaciation and deglaciation and cannot possibly be a coincidence (Harland & Rudwick, 1964; Walker, 2003). According to them, something must have triggered the speciation---the equivalent of the Big Bang as far as speciation (it is almost certain that there is a gap between the Ediacaran fossils and the Cambrian fossils which, when found, may answer some questions (Kerr, 1995(b); Palmer, 1996)). There is an iridium anomaly at the Cambrian-Precambrian boundary and a massive impact could have precipitated the melting of Snowball Earth, particularly if the Earth's crust was shattered (and some scientists are of the opinion that volcanism may have contributed to the melting). Others have postulated other reasons for organisms evolving (Mayell, 2001; Parker, 2003). Hoffman and Schrag (2002) have postulated that deglaciation was abrupt, followed (for reasons that we cannot go into detail here) by extreme heating. Additionally, Nursall (cf. Knoll, 2003) pointed out that the onset of oxygenation in the planet resulted in metazoan physiology, noting that the abundance and diversity of animal species declines sharply as we approach anoxic waters. Keep these postulates in mind as you evaluate the McClintock Effect.

And this is as good a place as any to bring up Parker's (2003) Light Switch Theory, as he calls it. Parker has pointed out in detail how important the sense of vision is, more so than any of the other senses, and, that the evolution of the complexity of vision from simple light receptors has not been accompanied by a corresponding sophistication in the other senses. 600 million years ago, there were no animals with eyes. 544 million years ago, there were no animals with eyes. Suddenly, "in the blink of an eye," as he likes to put it, 543 million years ago, there were animals with eyes. At the beginning of the Cambrian there was a hiatus of speciation, after which things calm down. The presence of vision in the first organism (a trilobite) was a boon for an organism which was a predator which was also prey. Vision, he claims, created "selective pressures" in other organisms that were either predators and/or prey so *they* evolved vision in the ensuing arms race. Unfortunately, Parker confuses cause and effect. The development of vision---in every animal that did so---was the effect. Vision in one animal did not "cause" other animals to "compensate" for their blindness by developing their own vision in an "arms race." If a predator had developed vision, which was indeed a monumental advantage, then that predator would have wiped out all other organisms and become supreme, followed by a population crash. Remember what happened to the dodo when an efficient predator was introduced into its ecology. He does make a good point, however, in pointing out that there was a thirty two million year gap between Snowball Earth and the Cambrian Explosion.

Snowball Earth aside, both the genesis and demise of ice ages are a mystery (glaciation is only one of many geological mysteries, along with the question of why do tectonic plates occasionally change direction of movement (Stone, 1995)). For mass glaciation to take place and to recede, several hypothesis have been advanced from the very beginning (Wallace, 1881/1998): the shifting shape of the continents as well as their topography, deep ocean currents (for example, when the Americas were joined at Panamá, the weather definitely changed, creating the Gulf Stream), eccentricities and "wobbles" in Earth's orbit (in 2003, it may be remembered, Mars made one of its closest perigees to the sun; it was strange seeing the planet without its ice caps), the level of solar energy reaching the planet, the level of carbon dioxide. Climate changes have taken place continually throughout the planet's history and, at present, a global warming is taking place whose prognosis and etiology is still being hotly debated, with political implications (Billups, 2004; Weissert & Bernasconi, 2004; Francois, 2004; Schiermeier, 2004; Moritz, Bitz & Steig, 2004; Coxall, Trathan & Murphy, 2004; Domine & Shepson, 2004; Johnson & Murphy, 2004). Of some relevance is the shape of the planet, strangely enough; the physical shape of the planet and its gravity field has changed; Earth is now more oblate, which may possibly affect its "wobble" (Fukumori, 2002).

The Great Ordovician Biodiversification Event

Schmitz, Tassinari & Peucker-Ehrenbrink (2001) and Schmitz, et al., (2008) have put the proposition that the Great Ordovician Biodiversification Event coincided with the asteroid belt being disrupted whereupon the Earth sustained numerous 1-km meteor impacts and resulted in extinctions Berry, Ripperdan & Finney, 2002)

The McClintock Effect

It is one of the premises of this book that not only are extraterrestrial impacts by meteors and comets---with the resulting volcanism, tsunamis, poisonous atmosphere, radical climactic alterations and temporary nocturnalization of the planet---have resulted in mass extinction of species, but they have also *simultaneously* caused speciation, without the benefit of Natural Selection, through what I call the McClintock Effect (basaltic traps/volcanism have been temporarily correlated with mass extinctions; extraterrestrial impacts have been temporarily correlated with mass extinctions; the conclusion is obvious).

It used to be thought that the genetic structure of an organism was a rigid one, with the genetic units held firmly in linear place. The usual image was of a string of beads on a string. Thanks primarily to the work of McClintock on maize, it is now known that there is a surprisingly degree of fluidity of these units and that the transposition of genetic elements may play a crucial role in an organism's genetic makeup (McClintock 1971; Keller, 1983; Nelson & Klein, 1984; Fedoroff & Botstein, 1992). Transposition, or translocation, involves the release of a genetic element from its loci and its insertion in another loci (Srb, et. al., 1965; Gardner, 1975). Minor "mutations" were due to a repositioning of elements within the chromosome, so that the function of a gene will vary with its position.[9]

In McClintock's *Spm* (suppressor-mutator) system, there are two components: the first, a "receptor," inserts into a gene, causing a phenotype, the second, more remote "regulator," controls the activity of the first. They become activated during periods of internal and/or external stress. She succinctly pointed out the evolutionary implications:

> The extraordinary range in types and times of control of gene expression during development that CEs [controlling elements] can elicit suggests to me that the basic mechanism may not be unduly diverse or complex. Indeed, variations of some common mechanism may be responsible for providing the many observed possibilities for control and integration of gene expression.

(McClintock, 1980; p. 16)

And more to the point:

> There is little doubt that genomes of some if not all organisms are fragile and that drastic changes may occur at rapid rates. These can lead to new genomic organizations and modified controls of type and time of gene expression. It is reasonable

> to believe that such genome shocks are responsible for the release of otherwise silent elements, which can then initiate changes to overcome disruptive challenges. I have emphasized that stabilizations do occur after release and action of maize CEs. Since the types of genome restructuring induced by such elements know few limits, their extensive release, followed by stabilization could give rise to new species or even new genera. (McClintock, 1980; p. 17)

Incredibly, no one seems to truly have grasped the ultimate deep significance of what she was saying, even when she repeated it, again and again (which was the case with much of her work, by the way):

> A few of these will be considered, along with several examples from nature implying that rapid reorganizations of genomes may underlie some species formation. (Mcclintock, 1984; p. 794)

And:

> I believe there is little reason to question the presence of innate systems that are able to restructure a genome. It is now necessary to learn of these systems and to determine why many of them are quiescent and remain so over very long periods of time only to be triggered into action by forms of stress, the consequences of which vary according to the nature of the challenge to be met. (Fedoroff & Botstein, 1992, p. 204)

And even in her Nobel lecture:

> The examples chosen illustrate the importance of stress in instigating genome modification by mobilizing available cell mechanisms that can

restructure genomes, and in quite different ways. A few illustrations from nature are included because they support the conclusion that stress, and the genome's reaction to it, may underlie many species formations. (McClintock, 1984)

From plants to animals (McClintock, 1950; 1961; 1984; Weiner, 1994). Stress proteins, also called "chaperones," have been found in all of the organisms that have been genetically studied and they help to protect proteins from environmental stress (e.g., oxygen starvation). One of these chaperones is HSP90 (Heat shock protein 90); it regulates developmental genes during times of stress by releasing previously hidden or buffeted phenotypic variance. And in the fruit fly it was found that heat shock (and presumably other stresses) increases the rate at which stress damaged proteins are reactivated, i.e., HSP90's protective role is neutralized, resulting in a different morphology, so that increased environmental variability results in an increase in variety within the organism (Cossins, 1998; Rutherford & Lindquist, 1998). What is doubly interesting is that epigenetics comes into play. In one experiment with fruit flies, even after normal HSP 90 activity was restored, the previous variation (legs growing out of eyes) persisted in a heritable manner through ten generations (Rutherford & Lindquist, 1998). In another fascinating study, it was found that when Mexican tetras were treated with an inhibitor of HSP90, they became eyeless, whereupon further investigation it was found that in caves, the stressor was the sharp drop in salinity of the water compared to the surface, thereby producing the hitherto puzzling blind cave fish (Pennisi, 2013; Rohner, et al.,

2013).

However, although there are obvious, consistent, genetic similarities in all organisms, by the same token there are differences, particularly between metazoans and bacteria, so that, although regulatory mechanisms in bacteria are known to respond to environmental stimulation, the attempt to obtain "stress-triggered evolution" through transposition has, so far, yielded negative results (Syvanen, 1988),[10] although Stindl (2014) and Shapiro (1992, 1999, 2005) are of the opinion that changes in phenotypes are due to transposon-mediated genomic repatterning. New technical advances may yield additional light on the subject (Jablonski, 2010; Levine, et al., 2014).

Extinctions

Mass extinctions have been shown to occur on an almost periodic basis of 26 million years and esoteric theories have been formulated to account for the cause of these extinctions, all of them being astronomical ones (Raup, 1999). Additionally, it has also been hypothesized that impact craters occur periodically, every 28 million years.

Whenever a cataclysm has occurred, whether it is global, or localized, we invariably see an extinction taking place, followed by speciation (it must be remembered that certain plant and animal species are very localized whereas others are widespread, so that a local catastrophe can destroy a local species). It is almost certain that the speciation is simultaneous, not subsequent to, with the extinctions; for example, we see a major radiation of angiosperms and pollinating insects right at the K-T boundary (Crepet, 1984).[11] Hitherto, it had been thought that extinctions were simply the opportunity for existing, surviving, species to fill in the vacuum, as it were, in taking over all of the vacated niches, whereupon evolution, i.e., speciation, took place, courtesy of Natural Selection (Courtillot & Gaudemer, 1996; Jablonski, 2004). Others (Hedges, Parker, Sibley & Kumar, 1996) have suggested that species diversification, at least in regards to the K-T boundary, began to occur prior to the extinction taking place, caused by the continental fragmentation that was taking place, presumably through Natural Selection.

I propose that the event that causes the extinction is itself the agent of speciation, *sans* Natural Selection.[12] Mass extinctions have been attributed to extraterrestrial impacts and massive volcanism; they have also sparked speciation. The ultimate stressor that induces speciation is astronomical/geological, affecting embryonic organisms (it may be that the actual catalyst is the isotopes that they carry and which saturate the global atmosphere, or just a particular area of the planet; some of the particular isotopes may be preferentially absorbed by some organisms), with epigenetic and symbiotic activities coming into play along with the McClintock Effect.

If the event is relatively localized, only a few species are affected (it is also important to emphasize that *groups* of individuals are mutated, not individuals). If on a continental, or global, scale, even whole Orders may be affected, which is why we see both allopatric and sympatric evolution. Nor should we rule out the role played by solar radiation whenever the Earth's magnetic field weakens and changes polarity, or whenever the sun's energy output varies (Kanipe, Talcott & Burnham, 1988), although, again, the overwhelming evidence is for a bolide impaction. Regardless, at present it is an unknown factor that which causes speciation and it is a rarely occurring one, which is why that is the real reason that we do not see widespread speciation continually taking place. Its presence is not constant, i.e., it is punctuated evolution.[13]

Note also that certain species have defied transformation: horseshoe crabs, Venus' Flytraps and coelacanths, just to name three. They seem to be very resistant to alterations in their genotype, possibly due to some yet to be discovered genetical "armor." Other species, on the other hand, are very susceptible to alteration: *Nepenthes* pitcher plants (95 species), *Drosera* sundews (90 species), *Anolis* lizards (143 species), cichlids (hundreds), tardigrades (1500 species) beetles (thousands of species), seed-shrimp (40,000 species).

Therefore, evolution is exogenous, not endogenous. Evolution is *imposed* on organisms. Evolution, i.e., phenotypic alteration, is *imposed* upon the organism. And it makes no difference that the organism is already very well adapted to its environment.

When a catastrophe does takes place, epigenesis, heterochrony, symbiosis and the McClintock Effect are activated. When this happens, a reshuffling of some of the gene expression, almost always dealing with the outer morphology, rarely the internal biochemical processes, takes place, so that speciation occurs. Some of the new traits may be incidentals (irrelevant to adaptation to the environment), others essential. Some phenotypes will be suppressed, others expressed, others reshuffled, others innovated. Note that this takes effect upon a group of organisms, not on an individual organism; apparently, the members of that group of incipient species' genotype is particularly pliable in a certain locus so that they are all altered along the same pattern. Note also that when this occurs the new traits may be maladaptive and the organisms may die shortly afterwards (McClintock, 1978), possibly by finding themselves in the wrong environment. "When we view the history of life from its beginnings some three billion years ago, we seem to witness a pattern of comparatively sudden revolutions interspersed with long intervals of relative quiescence." (Vermeij, 1977; p. 245)

Additionally, we get convergent evolution when suppressed genotype clusters, upon the advent of a new stressor will become reactivated after having been passed along across genera and even orders in a dormant state (keep in mind that a substantial portion of the genetic information is supposedly "dormant;" in humans it is 95%), which is inappropriately referred to as "junk."[14] If this seems a little fantastic consider the fact that it is generally acknowledged that the eye did not develop independently in 40 different branches of the evolutionary tree, but that the regulatory gene, *Pax 6,* is found to be involved in all animals that have eyes, and, that the same gene in different organisms may have a different phenotypic expression (Mayr, 2001). As McClintock (1984) wrote, in what is most certainly an understatement: "It is becoming increasingly apparent that we know little of the potentials of a genome. Nevertheless, much evidence tells us that they must be vast." (p. 795)

It is also important to take note of the exhaustive work carried out by Schindewolf (1950/1994), Eldredge and Gould (1972), Kaufmann (Fortey, 2000), Hunt (2007), and others (Jablonski, 2010), using massive amounts of data, that have conclusively proven that initially after a mass extinction there is a burst of speciation, followed by a decline and stasis, until the next cataclysm occurs, mass extinction follows and the cycle gets repeated again and again throughout millions of years.

Some researchers have postulated that there were no sudden mass extinctions since the fossil record indicates a gradual decline in specimens. However, others have attributed this decline to sampling bias rather than a decline in terrestrial diversity whereas oceanic extinction is abrupt (Signor & Lipps, 1982).

One of the conundrums in paleontology is accounting for the sudden Avalon Explosion of the soft-bodied Ediacara biota after the end of the Cryogenian period and later of the explosion of the hard-bodied species in the Cambrian. A factor may have been the Great Oxidation Event, the increasingly prevalent presence of a poisonous gas in the atmosphere---oxygen. As Carroll (2000) has pointed out, the physical history of our planet may have gone hand in hand with evolutionary processes.

In conclusion, mass extinctions occur upon an impact by a meteor(s) or comet(s) big enough and fast enough to shatter the ultrathin planetary crust, whereupon flood basaltic eruptions cascade outwards, resulting in noxious gasses, extreme temperatures, acidification and sulfuric poisoning, aside from the direct impact emanations of shock wave, iridium radiation, extreme heat, sonic blast and the altered climate. Global or semi-global catastrophes create an environment that is exceedingly, unimaginably, stressful (Keller, Punekar & Mateo, 2015; Font, et al., 2016). Our mayfly worldview cannot begin to realistically imagine it. Contrary to what has been traditionally assumed, such a crisis does not offer an opportunity for evolution to take place through the gradual process of Natural Selection, in so far as animals taking advantage of newly available niches, but rather a period of hiatus takes place wherein the genome of many organisms is directly altered by the trauma and the altered environment itself. If these cataclysmic stresses are multigenerational, which they must be, the epigenetic and symbiogenetic changes become permanent. Some organisms are highly resistant to change; others are highly plastic (and this difference may be found within the same species, so that you will have speciation while retaining the original species). Some of the new organisms thrive as a result of the metamorphosis while others die right away while still other, original species survive in microclimates far from the cataclysm. Most importantly, the fossil record of punctuated evolution, and the paradox of the "living fossils" (in other words, the data) fits this hypothesis. In support of the above, Courtillot & Olson (2007) make an astoundingly obvious, yet overlooked, piece of evidence: "The record also includes evidence of the converse relationship, that traps are rare or missing during times without mass extinctions." (p.497)

To reiterate once more: every major extinction event has been accompanied by an astronomical impact, followed by flood basaltic eruption, resulting in extinction of species and a subsequent, sudden, diversification of new species. This diversification in phenotypes is due to the stress-induced alteration of the transposon-derived repetitive sequences, very much like the letters of the alphabet can be rearranged to form new words, as well as symbiogenesis, heterochrony and epigenesis.[15]

This hypothesis is a better fit than the Natural Selection hypothesis.

FOOTNOTES TO CHAPTER NINE

"Go and see."
---Nicolas Desmarest

Change automatically suggests vigor, plasticity and youth. In contrast, rigidity is associated with rest, cerebral lassitude, and paralysis of thought; in other words, fatal inertia---certain harbinger of decrepitude and death.
---Santiago Ramón y Cajal, *Advice for a Young Investigator*

No matter how many times Darwinists reiterate the fairy story that the homologs were "once upon a time" adaptive in the ancestor of the clade they define, it is a claim without the slightest empirical or rational basis.
---Richard Denton, *Evolution: Still a Theory in Crisis*

[1] It survived. By a miracle.

[2] There is some evidence that the poles might be getting ready to repeat the process, according to one computer model, which shows that a magnetic anomaly corresponding to about where the hole in the ozone layer is presently located should be taking place (and there is actually a magnetic anomaly at that spot). Furthermore, and much more importantly, the Earth's magnetic field is rapidly decreasing.

No one, of course, really knows why the magnetic poles have changed (Southwood, 2003).

However, others have suggested that the reversal of the Earth's magnetism is correlated, if not actually caused by, extraterrestrial impacts (Erickson, 1991). This makes sense if one considers that a magnet which is forcefully hit has its magnetism altered.

[3] One of the amusing things about giving the odds for unlikely events is that the events nonetheless come to pass. I am sure that the reader has been warned many times against buying a lottery ticket because the odds against winning are so many millions to one against. Yet, every month, and sometimes, every week, a winner does win the lottery. In fact, I have an acquaintance who won seven million dollars in the lottery. And I was recently in a room with seven other random strangers. It turned out that three of those individuals had identical twin siblings.

[4] Wignall & Pritchett (1996) have accumulated data of rapid oceanic anoxia at the end of the Permian period. Animals that survived may have done so because either they found themselves in beneficial microclimates, or because their oxygen requirements were low.

[5] Mars' surface, incidentally, also shows numerous impacts, but it has been theorized, with much evidence, that it once had seas (and sandstorms do occur, visible from Earth with a telescope) which would mean that the impacts took place late in its history since erosion would have weathered away the craters. This is no surprise, considering Mars' proximity to the asteroid belt (Albee, 2004). Its two "moons," in fact, are just two small asteroids that got trapped in its orbit, instead of impacting in the surface.

[6] Alvarez (1998) makes the point that with the

numerous space probes to planets and moons and the landings on the moon, which demonstrated that impact craters are the rule, not the exception, geology should have been transformed into an interplanetary, catastrophic, discipline. Instead, this transformation was derailed by an equally dramatic, contemporary, discovery which gave the uniformitarian viewpoint a new lease on life---gradual, grinding, plate tectonics and continental drift. Earlier, Darwin's gradual evolution echoed Lyell's view of geology.

[7]It is generally believed that if a meteor or comet was detected on a collision course with Earth, we would send up a rocket with a nuclear bomb which would explode near the target, momentarily vaporizing the object (the fragments would rematerialize again once the heat dissipated) and presumably render it harmless. No test runs have been made by either Russia, Europe, or America to test that hypothesis with any of the known orbiting asteroids on a noncollision near earth orbit, something that has been urged, without success, since 1981 (James, 1981). It is also forgotten that the blast of an atomic explosion occurs in an atmosphere. There is no atmosphere in space.

[8]I have kept track over the years, in an informal manner, of the *zeitgeist* regarding a possible extraterrestrial impact ever since, as a teenager, I read in a small article in *Science Digest,* sometime in the 1960s, about a near miss with the Icarus asteroid.

In the 1940s, a classic of science fiction literature came out, entitled *When Worlds Collide.* It depicted the end of the world through the collision of a large planet with our own; a smaller planet circled the larger one and it settled around the sun around the same orbit as Earth, so that the story revolved about humans building a ship to transfer themselves over to the new world. Among its original elements, it introduced the need for preserving and transferring our ecology and not just human beings. A sequel, *After Worlds Collide,* was less interesting, as was *The Big Eye* by Max Erhlich about another planet passing by Earth.

The 1950s saw the advent of Immanuel Velikovsky with his *When Worlds Collide* (I have to admit that I was never able to get past the first chapter). Velikovsky was the boogey man of uniformitarians and gradualists since he seriously proposed that Venus had been spit out of Jupiter in the form of a comet and that the planets Mars and Venus had come close to Earth and caused untold havoc and that such encounters had been preserved in human legends. To say that scientists responded in a hostile, emotional, manner is an understatement. To say that Velikovsky's science was shoddy is also an understatement. At any rate, he served to entrench uniformitarians (Goldsmith, 1977).

Anyway, after that *Science Digest* 1960s article, there were occasional one-inch newspaper stories, or magazine notices, that a near collision had occurred recently. Very rarely was the subject of extraterrestrial collisions broached (e.g., Rubinsky & Wiseman, 1982). It was not until the 1980s, when the Alvarez team announced their findings and magazine articles and TV shows began coming out with dramatic details of what probably immediately happened as a result of the K-T impact, that people started to take notice and ask questions as to whether it could happen again. In 1992,

Brian Marsden announced that Comet Swift-Tuttle, fifteen kilometers in diameter, would slam into Earth on August 2126, which made headline news; months later, the comet's orbit was revised (Levy, 1995). Then, Comet Shoemaker-Levy 9 completely changed public perception. In 1998, Marsden was once again involved in another false alarm, albeit peripherally, when it was announced that asteroid 1997 XF11 was going to intercept the planet in 2028 (Reichhardt, 1998). Then, Hollywood stepped in. A very good movie, *Deep Impact,* fictionalized in 1998 just such a future possibility. It was immediately followed by an abysmal movie, *Armageddon,* the kind of film that is so awful, you find yourself cringing inside the movie theater, gritting your teeth and clawing the armrests of your chair. This was followed by an even worse---if such a thing was possible---TV miniseries over the very same topic whose title I have thankfully forgotten. Since then, the subject has come up numerous times in television cartoons and in men's magazines (Liddell, 2004), of all things, with the stated message that if a meteor ever gets through, we are all finished. Regardless, the increased interest in potential impacts has resulted in the tracking of 200 presently known asteroids that cross Earth's orbit, whereas in 1972 only a dozen were known (Levy, 1995). NASA, as a result of Comet Shoemaker-Levy 9, was forced to begin scanning the system for possible impact asteroids, though the bureaucracy was initially resistant (Kaiser, 1995), but it later toed the line (Roach, 2003); as early as 1981, as a result of the Alvarez team's findings, Eugene Shoemaker had been urging NASA, without success, to establish Project Spacewatch (James, 1981).

Incidentally, as late as the 1960s, the 1908 explosion over Tunguska used to be grouped with pseudoscientific topics, such as the Bermuda Triangle and the Loch Ness Monster.

[9] A telling reminiscence of McClintock (Green, 1992, p.119): During the discussion, someone put the question to Sturtevant:

> "Sturt, what did McClintock have to say?" Sturt sucked on his ever-present pipe, looked briefly at the ceiling, and then replied. "I didn't understand one word she said, but if she says it is so, it must be so!" Such was the reputation and confidence in Barbara widespread among the maize and *Drosophila* geneticists of the period.

[10] One pertinent fact is that in early studies it was thought that insertion sequences of genetic material became inserted in random loci; later studies have shown that many elements have preferential target sites (Syvanen, 1988).

[11] When "speciation" take place, it is usually through allopolyploidy via hybridization---and almost always in plants---but, as Coyne (1996) has pointed out, this is very rare, otherwise we would see a profusion of hybrid swarms to the point that we would be up to our necks in hybrids. Nor, I might add, have we seen speciation taking place through Natural Selection or in connection with volcanism, so by a process of elimination we can eliminate them from our list of suspects.

[13] Sill (2004) points to the significant effect of environmental stress in marginal habitats as a mechanism of rapid evolutionary change, particularly

when the dinosaurs displaced the therapsids, but, if he doubts the role of Natural Selection he does not say so. Although he confines himself to the late Triassic, if one looks closely, it is suggestive that paleontological eras tend often to be associated with impact craters, glaciation and/or mass extinctions. Again and again we see widespread extinctions associated with mass speciation and catastrophes. Knoll (2003) makes a similar observation (p.223): "If there is one lesson that paleontology offers to evolutionary biology, other than the documentation of biological history itself, it is that life's opportunities and catastrophes are tied to Earth's environmental history."

Of relevance here is the finding by Sepkoski and Raup that mass extinctions show periodicity; Sepkoski emphasized that the periodicity by itself does not explain the cause of the mass extinctions, nor whether they were gradual or sudden (Briggs, 1999). An astronomical factor was speculated.

[14]If this theory is correct, then, conceivably, extinct species could be brought back in the future if the specific genetic clusters that have become dormant could be identified and artificially reactivated in a sophisticated manner, so that, for example, horseshoe crabs or chitons could be made to yield a species of trilobite. This is pure speculation, of course, pure fantasy with just a hint of possibility. However, if this fantasy turns out to have a basis in fact, then it brings to mind the nineteenth century German biologists' assertion regarding embryogenesis (Richards, 1992). Perhaps Haeckel's famous dictum that ontogeny recapitulates phylogeny may be more than just a pithy, mistaken,

epigram.

[15]Wei, et, al., (2014) have put forth an intriguing hypothesis to wit that geomagnetic reversals have resulted in the loss of O^+ into space, thereby causing mass asphyxiation. Perhaps it is attributing too much to impacts, but a question that immediately arises is whether a very large astronomical impact could cause a geomagnetic reversal.

CONCLUSION

Good science is never static.
---Conrad Zirkle, *Evolution, Marxian Biology and the Social Scene*

Evolution can be an experimental as well as an observational science.
---Richard Lenski

False facts are highly injurious to the progress of science for they often endure long: but false views, if supported by some evidence, do little harm, for everyone takes a salutary pleasure in proving their falseness.
---Charles Darwin

The field of evolution has essentially remained stagnant since Darwin and Wallace first proposed their theory. No real progress has been made. Evolution just . . . happens. Like spontaneous generation. One simply has to wait a long time to see it. It cannot be tested. It cannot be experimented upon. It cannot be proven or disproven. It is as if the state of science had become frozen since Victorian times.

Proof of what I am saying can be seen in comparing the debate on evolution with other scientific subjects, such as the research on ovarian cancer, cladistics, HIV, or plate tectonics. In those areas, highly technical data is cited during discussions, using specialized technical language. With evolution, anyone can just walk in and give his/her opinion. Indeed, as I previously stated, philosophers, journalists, lawyers and theologians, who specialize in verbosity and nothing else have written books on the subject and given lectures, which I have unfortunately attended.

The writings of some scientists trying to reconstruct past events, particularly biologists, paleontologists and anthropologists, are full of "would haves," "could haves," "probablys," "should have beens," "could have beens," "almost certainly weres," "chances ares." It reminds me of a television cartoon character bridging a chasm by building a footpath of bricks as it progresses, with nothing but air under it supporting its architectural endeavor and oblivious to the absurdity of the situation. Typical culprits in this absurdity are Margulis and Sagan (1997) in *Microcosmos* and Dawkins (1982) in *The Extended Phenotype* where rampant speculation occurs throughout. They convince themselves that the hypothetical scenarios that they picture are, indeed, factual and historical and, to make matters even more absurd, they offer those same scenarios as proof of their theoretical stand (I have to admit that halfway through *The Extended Phenotype,* in particular, I found myself grinding my teeth).[1]

I am usually an experimenter, not a theoretician. As a rule, I dislike new theories. I will take facts over theory any day of the week. Nonetheless, I have put forth in this book a theory of a mechanism for evolution which I am fairly confident is accurate. More importantly, the details can be proven or disproven empirically, something that has been lacking in the case of Natural Selection which put forth the teleological idea of Nature as a selective breeder. And, at no point did I state the neo-Darwinists' favorite phrase: "We must assume that--"

Prove me wrong. But please do not give us computer programs that write little squiggles that look like bugs, or clams, and call it evidence of Natural Selection. Nor give us arguments containing a long list of assumptions of what may have happened with fauna and flora millions of years ago. Nor give us arguments employing analogies. Give us facts. Give us data. Give us experiments. And, please, no sentient plants and animals poring over nonexistent blueprints.

If nothing else, this theory will be considered a monumental success by me if, at least, it helps to end the stagnation once and for all. If the present theory put forth in this book should be proven to be wrong---and it may just turn out to be total rubbish from beginning to end---then toss this book in the nearest trash can! But, remember that the basic problem remains. Because that is all that a theory should be: a tool. That is all. A way of organizing data and a source of prediction. And if a particular theory cannot do the job, then amend it, or, get another one that will. A theory should most definitely not be an object of unquestioned reverence, no matter how elegant, or how beautifully stated.

The justification of all mathematical models is that, oversimplified, unrealistic, and even false as they may be in some respect, they force analysts to confront possibilities that would not have occurred to them otherwise. The history of physics and medicine abounds with wrong or incomplete theories that throw just enough light to allow some other big breakthroughs. The atom bomb, for example, was built before physicists understood the structure of particles. (Nasar 1998, p.120)

It has often been said in science that no matter how many facts may support a theory, it only takes one to negate it. If this is so, then the classical theory of evolution should have been negated long ago, or more accurately amended, as Newton's laws were amended with Einstein's theories. This is not to say that the concept of evolution itself should be eliminated, of course; the evidence for evolution is overwhelming, but, rather, that the *mechanism* that brings about evolution needs to be revised. The classical Darwinian-Wallace theory was a nineteenth century theory, by which I mean that it was a simplistic theory, as simplistic as the old, pre-McClintock view of the genome as a static repository of genetic information (Federoff & Botstein, 1992), or of genes as a string of beads within the chromosome. Indeed, its simplicity was both its appeal and its shortcoming. The reality was much more complicated. Evolution of the species is a much more complicated business than nineteenth century science could possibly have accounted for. This is not to say that we have all the answers. We have just barely scratched the surface and seen that, underneath, it is very, very complex.

But one thing is certain: Natural Selection is no longer a viable mechanism for speciation, though it is for microevolution. The data overwhelmingly supports this conclusion. British and American neo-Darwinists, however, with their tunnel vision and their arrogance, insist that nothing at all is wrong and have raised the theory to the level uninhabited by any other theory, to the level of sacred dogma,[2] thereby tarnishing an otherwise beautiful theory, along with also tarnishing the names of Darwin and Wallace. They have retarded progress in the field by insisting that the classical theory is perfect as it is and should not be questioned, should not be altered in any way, and anyone who attempts to do so must be a Creationist and therefore should be ignored altogether without so much as looking at the conflicting data. It may be remembered that these neo-Darwinists called for a boycott of *New Scientist* by scientists simply for picturing the tree of interrelated species as a web rather than a tree. We cannot continue to waste our time with these people, holding up the progress of science, by trying to convince people who are rigidly close minded. We simply have to bypass them and get on with the job of science.

Finally, I will end with just one last quotation---for this chapter, that is (MacArthur & Wilson, 1967, p. 5):

> A good theory points to possible facts and relationships in the real world that would otherwise remain hidden and thus stimulates new forms of empirical research. Even a first, crude theory can have these virtues. If it can also account for, say, 85% of the variation in some phenomenon of interest, it will have served its purpose well.

FOOTNOTES TO CONCLUSION

The scientist must not forget that hypotheses are a means, never an end.
---T. H. Huxley

But if we face the Burgess fauna honestly, we must admit that we have no evidence whatsoever---not a shred---that losers in the great decimation were systematically inferior in adaptive design to those that survived.
---S. J. Gould, *Wonderful Life*

It is a proof of the great fact that I have already maintained in public, that several animal species have been entirely destroyed by the revolutions that our planet has undergone.
---Georges Cuvier

[1]Max Planck is reputed to have stated that science advances one funeral at a time: "A new scientific truth does not triumph by convincing its opponents and making them see the light, but rather because its opponents eventually die, and a new generation grows up that is familiar with it." (Barber, 1960; p.38). Or, as Asimov paraphrased it, "Science advances one funeral at a time." Dawkins' funeral is eagerly anticipated.

[2]Ironically, their stance is very similar to those religious persons' faith: one can point to the inherent contradiction of believing in an all-powerful, omniscient

Deity who is obsessed with Good and with the triumph of Good over Evil on the one hand and on the other hand, the fact that there is so much Evil in the world that is rampant and triumphant. All sorts of rationale are given to explain away the obvious contradiction (at least in the Catholic faith; with Protestants and Muslims, the dilemma is simply ignored, or suppressed).

Likewise, those neo-Darwinists who preach that the completely unaltered uniformity of morphology in many species over millions of years attests to the slow process of Natural Selection while, at the same time, welcoming with Hosannas current studies showing that some organism's morphology became altered by a fraction of a centimeter after five years because of a changed environment are identical to evangelical preachers who assure their parishioners that God always answers all prayers then afterwards have to console those same parishioners because God did not change their circumstances, or, to the feminist lawyers who argue that women should be admitted in male-only professions because women are no different than men and should not be treated differently while simultaneously insisting that the jobs make physical changes in the working environment in order to accommodate those same women.

REFERENCES

Adler, Jerry & Carey, John. Is man a subtle accident? *Newsweek,* 3 November 1980, 95-96.

Adriaens, Pieter. Evolutionary psychiatry and the schizophrenia paradox: a critique. *Biology and Philosophy,* 2007, *22,* 513–528.

Aggarwal, Ramesh. Ancient frog could spearhead conservation efforts. *Nature,* 2004, 1 April 2004, *428,* 467.

Akst, Jef. (2017) Evolution's quick pace affects ecosystem dynamics. *The Scientist.* http://www.the-scientist.com/?articles.view/articleNo/49258/title/Evolution-s-Quick-Pace-Affects-Ecosystem-Dynamics/

Albee, Arden. The unearthly landscapes of Mars. *Scientific American,* June 2003, *288,* 44-53.

Alcorn, Gordon. *Owls.* New York: Prentice Hall Press, 1986.

Alexander, George. Going, going, gone. *Science 81,* May 1981, 65-69.

Alfven, Hannes. Plasma physics, space research, and the origin of the solar system. *Science,* 4 June 1971, *172,* 991-995.

Allen, David. *The Naturalist in Britain.* Princeton: Princeton University Press, 1994.

Altemeier, William. Bugs in the OR. *Emergency Medicine,* August 1974, 95-110.

Alvarez, Walter. *T. rex and the Crater of Doom.* Princeton: Princeton University Press, 1997.

---------Claeys, Philippe & Kieffer, Susan.

Emplacement of Cretaceous-Tertiary boundary shocked quartz from Chicxulub crater. *Science,* 18 August 1995, *269,* 930-935.

Allen, Andrew; Brown, James & Gillooly, James. Global biodiversity, biochemical kinetics, and the Energetic-Equivalence Rule. *Science,* 30 August 2002, *297,* 1545-1548.

Andrews, Paul & Thompson, J. Anderson. The bright side of being blue: Depression as an adaptation for analyzing complex problems. *Psychological Review,* 2009, *116,* 620–654.

Anway, M. D.; Cupp, A. S.; Uzumcu, M. & Skinner, M. K. Epigenetic transgenerational actions of endocrine disruptors and male fertility. *Science,* 2005, *308,* 1466-1469.

Anonymous. Mirror sight. *Nature,* 29 April 1999, *398,* ix.

Archibald, J. David. No, it only finished them off. *Natural History,* May 2005, *114,* 52-53.

Artemieva, Natalia; Morgan, Joanna & Expedition 364 Science Party. Quantifying the release of climate-active gases by large meteorite impacts with a case study of Chicxulub. *Geophysical Research Letters,* 2017, http://onlinelibrary.wiley.com/doi/ 10.1002/2017GL074879/full?hootPostID=ed358 a4dd6182fffcee18bdd66f79761

Astor, Michael. Scientists claim to discover new fish. http://aolsvc.news.aol.com/news/ article.adp?id=20040618014409990004.

Attenborough, David. *The Private Life of Plants.* Princeton: Princeton University Press, 1995.

Bakker, Robert. *The Dinosaur Heresies.* New York: Zebra Books, 1986.

Ball, Philip. How natural is numeracy? *Aeon,* 2017, https://aeon.co/essays/why-do-humans-have-numbers-are-they-cultural-or-innate.

Balter, Michael. Brain man makes waves with claims of recent human evolution. *Science,* 22 December 2006, *314,* 1871-1873.

Barber, Bernard. Resistance by scientists to scientific discovery. *Scientific Manpower,* 1960, 36-47.

Bard, Edouard; Hamelin, Bruno; Arnold, Maurice; Montaggioni, Lucien; Cabioch, Guy; Faure, Gerard & Rougerie, Francis. Deglacial sea-level record from Tahiti corals and the timing of global meltwater discharge. *Nature,* 18 July 1996, *382,* 241-244.

Barthlott, Wilhelm; Porembsk, Stefan; Seine, Rüdiger & Theisen, Inge. *The Curious World of Carnivorous Plants.* London: Timber Press, 2007.

Baxter, John. *Science Fiction in the Cinema.* London: Tantivy Press, 1974.

Becker, Luann; Poreda, Robert J.; Hunt, Andrew G.; Bunch, Theodore E. & Rampino, Michael. Impact event at the Permian-Triassic boundary: Evidence from extraterrestrial noble gases in fullerenes. *Science,* 2001, *291,* 1530-1533.

Becker, L.; Poreda, R.; Basu, A; Pope, K.; Harrison, T.; Cholson, C. & Lasky, R. Bedout: a possible End-Permian impact crater offshore of Northwestern Australia. *Science,* 2004, *304,* 140-146.

Behe, Michael. *Darwin's Black Box.* New York: Simon & Schuster, 1996.

----------*The Edge of Evolution.* New York: Free Press, 2007.

Benton, Michael. Dinosaurs' lucky break. *Natural History,* June 1984, *93,* 54-59.

---------*When Life Nearly Died.* London: Thames & Hudson, 2003.

Benz, Francis. *Pasteur: Knight of the Lab.* New York: Dodd, Mead & Co., 1938.

Berry, William; Ripperdan. Robert & Finney, Stanley. Late Ordovician extinction: A Laurentian view. In Koeberl, Christian and MacLeod (Ed.) *Catastrophic Events and Mass Extinctions.* Boulder: Geological Society of America, 2002.

Billups, Katharina. Low-down on a rhythmic high. *Nature,* 19 February 2004, 686-687.

Bird, W. R. *The Origin of Species Revisited.* New York: Philosophical Library, 1989.

Blease, C. R. Too many 'Friends,' too few 'Likes'? Evolutionary psychology and 'Facebook Depression.' *Review of General Psychology,* 2015, *19*, 1–13.

Bloch, Sidney & Reddaway, Peter. *Psychiatric Terror.* New York: Basic Books, 1977.

Bock, Ralph. The give-and-take of DNA: Horizontal gene transfer in plants. *Trends in Plant Science,* 2009, *15*, 11-22.

Bohr, Bruce & Seitz, Russell. Cuban K/T catastrophe. *Nature,* 12 April 1990, *344,* 593.

Boessenkool, Sanne; Taylor, Sabrina; Tepolt,

Carolyn; Komdeur, Jan & Jamieson, Ian. Large mainland populations of South Island robins retain greater genetic diversity than offshore island refuges. *Conservation Genetics,* 2007, *8,* 705-714.

Bowers, Faubion. *Broadway USSR.* Toronto: Thomas Nelson & Sons, 1959.

Bown, Matthew. *Art Under Stalin.* New York: Holmes & Meier, 1991.

Bradshaw, William & Holzapfel, Christina. Evolutionary response to rapid climate change. *Science,* 9 June 2006, *312,* 1477-1478.

Braun, David. Researchers rethink dinosaur die off scenario. http://news. national geographic.com/news/2002/02/0222_020222-dinodust.html.

Brett, C. &Baird, G. (1995) "Coordinated stasis and evolutionary ecology of Silurian to Middle Devonian faunas in the Appalachian Basin." In D. Erwin & R. Anstey (Eds.) *New Approaches to Speciation in the Fossil Record.* New York: Columbia University Press. Pp. 285-317.

Bretz, J Harlen. The Lake Missoula floods and the channeled scablands. *Journal of Geology,* 1969, *77,* 505-543.

Briggs, Derek. J. John Sepkoski Jr. (1948-1999). *Nature,* 5 August 1999, *400,* 514.

Brooks, Thomas & Balmford, Andrew. Atlantid forest extinctions. *Nature,* 14 March 1996, *380,* 115.

Broyles, Robyn. Punctuated Equilibrium. http://www.geocities.com/CapeCanaveral/Lab/1366/pe.html.

Burr, T.; Hyman, J. & Myers, G. The origin of acquired immune deficiency syndrome: Darwinian or Lamarckism? *Philosophical Transactions of the Royal Society of London Biological Science,* June 21, 2001, *356,* 877-87.

Butterfield, N. J. Bangiomorpha pubescens: Implications for the evolution of sex, multicellurarity, and the Mesoproterozoic/Neoproterozoic radiation of eukaryotes, *Paleobiology,* 2000, *26,* 386-404.

Carr, Edward. *What is History?* New York: Vintage Books, 1961.

Carroll, Robert. Towards a new evolutionary synthesis. *Tree,* 2000, *15,* 27-32.

Carroll, Sean. The origins of form. *Natural History,* November 2005, *114,* 58-63.

Cattell, Raymond,; Eber, H. & Tatsuoka, M. (1970) *Handbook for the Sixteen Personality Factor Questionnaire (16PF).* Champaign, Il.: IPAT.

Cattell, Raymond. *The Scientific Use of Factor Analysis.* New York: Plenum Press, 1978.

----------Superseding the Motivational Distortion scale. *Psychological Reports,* 1997, *70,* 499-502.

Chapman, Clark. Impact lethality and risks in today's world: Lessons for interpreting Earth history. In Koeberl, Christian and MacLeod (Ed.) *Catastrophic Events and Mass Extinctions.* Boulder: Geological Society of America, 2002.

Chapman, Tracey & Partridge, Linda. Sexual conflict as fuel for evolution. *Nature,* May 16, 1996, *381,* 189-190.

Chen, J.Y,; Dzik, J.; Edgecombe, G. D.; Ramskold, L. & Zhou, G.-Q. A possible Early Cambrian chordate.

Nature, 26 October 1995, *377,* 720-722.

Coalacino, Carmine. Leo Croizat's biogeography and macroevolution, or . . . "Out of nothing, nothing comes." *Philippine Scientist,* 1997, *34,* 73-88.

Collin, Rachel & Cipriani, Roberto. Dollo's law and the re-evolution of shell coiling. *Proceedings of the Royal Society of London B,* 22 December 2003, *270,* 2551-2555.

Conquest, Robert. *The Great Terror.* London: Macmillan & Co., 1968.

Conway Morris, Simon. *The Crucible of Creation.* Oxford: Oxford University Press, 1998.

Coon, Carleton. *The Living Races of Man.* New York: Alfred A. Knopf, 1970.

Cortijo, Sandra; Wardenaar, Rene; Colome-tatche, Maria; Gilly, Arthur; Etcheverry, Mathilde; Labadie, Karine; Caillieus, Erwann; Hospital, Frederic; Aury, Jean-Marc; Wincker, Patrick; Roudier; François; Jansen, ritser; Colot, Vincent & Johannes, Frank. Mapping the epigenetic basis of complex traits. *Science,* 2014, *343,* 1145-1148.

Cossins, Andrew. Cryptic clues revealed. *Nature,* 26 November 1998, *396,* 309-310.

Courtillot, Vincent & Gaudemer, Y. Effects of mass extinctions on biodiversity. *Nature,* 9 May 1996, *381,* 146-148.

---------& Olson, Peter. Mantle plumes link magnetic superchrons to Phanerozoic mass depletion events. *Earth and Planetary Science Letters,* 2007, *260,* 495-504.

Cox, Alexanders & Keller, Brenhin. (2023) A Bayesian

inversion for emissions and export productivity across the end-Cretaceous boundary. *Science, 381,* 1446-1451.

Coyne, Jerry. Not black and white. *Nature,* November 1998, *396,* 35-6.

---------Speciation in action. *Science,* 3 May 1966, *272,* 700-701.

---------& Charlesworth, Brian. Mechanisms of punctuated evolution. *Science,* 6 December 1996, *274,* 1748-1749.

Crepet, William. Ancient flowers for the faithful. *Natural History,* April 1984, *93,* 39-44.

Crooks, Kevin & Soule, Michael. Mesopredator release and avifaunal extinctions in a fragmented system. *Nature,* 5 August, 1999, *400,* 563-566.

Cropley, J. E.; Beckman, K. B. & Martin, D. Germ-line epigenetic modification of the AVY murine allele by nutritional supplementation. PANS, November 14, 2006, *103,* 17308-17312.

Croxall, J.; Trathan, P. & Murphy, E. Environmental change and Antarctic seabird populations. *Science,* 30 August 2002, *297,* 1510-1514.

Culotta, Elizabeth. Exploring biodiversity's benefits. *Science,* 23 August 1996, *273,* 1045-1046.

Cuvier, Georges. (1836) http://www.victorianweb.org/science/science_texts/cuvier/cuvier_onlamarck.htm

Czamanske, Gerald K. & Fedorenko, Valeri A. The Demise of the Siberian Plume, 2017. http://www.mantleplumes.org/Siberia.html

Darwin, Charles. *The Voyage of the Beagle.* New York:

Anchor Books, 1860/1962.

---------*The Origin of Species.* New York: Gramercy Books: 1859/1979.

---------*The Expression of the Emotions in Man and Animals.* London: John Murray, 1872.

Dawkins, Richard. *The Extended Phenotype.* Oxford: Oxford University Press, 1982

----------*The Blind Watchmaker.* London: W. W. Norton: 1986.

---------God's utility function. *Scientific American,* November 1995, *231,* 80-85.

Debre, Patrice. *Louis Pasteur.* Baltimore: Johns Hopkins University Press, 1994.

Dell, P. A. & Rose, F. D. transfer of effects from environmentally enriched and impoverished female rats to future offspring. *Physiology & Behavior,* 1987, *39,* 187-190.

Denton, Michael. *Evolution: Still a Theory in Crisis.* Seattle: Discovery Institute Press, 2016.

Diamond, Jared. Did Komodo dragons evolve to eat pygmy elephants? *Nature,* 30 April, 1987, 350-352.

---------Daisy gives an evolutionary answer. *Nature,* 14 March 1996, *380,* 103-104.

Dias, Brian & Ressler, Kerry J. (2014) Parental olfactory experience influences behavior and neural structure in subsequent generations. *Nature Neuroscience, 17,* 89–96.

Dietrich, Michael. (2000) From hopeful monsters to homeoetic effects: Richard Goldschmidt's integration of development, evolution, and genetics. *American Zoologist, 40,* 738-47.

---------(2003) Richard Goldschmidt: hopeful monsters and other "heresies." *Nature Review, 4,* 68-74.

Domine, Floretn & Shepson, Paul. Air-snow interactions and atmospheric chemistry. *Science,* 30 August 2992, 1506-1510.

Donnadieu, Yannick; Godderis, Yves; Ramstein, Gilles; Nedelec, Anne & Meert, Joseph. A 'snowball Earth' climate triggered by continental break-up through changes in   runoff. *Nature,* 18 March 2004, *48,* 303-306.

Dunn, Robert. Blurring Wallace's Line. *Natural History,* September 2004, *113,* 61.

Ehrenfeld, David. Vanishing knowledge. *Harper's Magazine,* March 1996, *187,* 15-17.

Eakin, C. Mark. Lamarck was partially right---and that is good for corals. *Science,* 2014, *344,* 798-799.

Eddy, Michael; Schoene, Blair; Samperton, Kyle; Keller, Gerta; Adatte, Thierry & Khadri, Syed.U-Pb zircon age constraints on the earliest eruptions of the Deccan Large Igneous Province, Malwa Plateau, India. Earth and Planetary Science Letters, 540, 116249.
https://doi.org/10.1016/j.epsl.2020.116249

Eisner, Thomas; Lubchenco, Jane; Wilson, Edward; Wilcove, David & Bean, Michael. Building a scientifically sound policy for protecting endangered species. *Nature,* 1 September, 1995, *268,* 1231-1232.

Eldredge, Niles. *The Triumph of Evolution and the Failure of Creationism.* New York: Henry Holt & Co., 2001.

----------& Gould, S. J. "Punctuated equilibria: an

alternative to phyletic gradualism," In T. Schopf (ed.). *Models of Paleobiology.* Dan Francisco: Freeman, Cooper & Co., 1972, pp. 82-115.

Elena, Santiago; Cooper, V & Lenski, R. Punctuated evolution caused by selection of rare beneficial mutations. *Science,* 21 June 1996, *272,* 1802-1804.

Ezenwa, Vanessa; Gerardo, Nicole; Inouye, David; Medina, Mónica & Xavier, Joao. Animal behavior and the microbiome. *Science, 338*,198-199.

Erdmann, Mark, Caldwell, Roy & Moosa, M. Kasim. Indonesian 'king of the sea' discovered. *Nature,* September 24, 1998, *395,* 335.

Erickson, Jon. *Target Earth!* Summit: TAB Books, 1991.

Erwin, Douglas. The mother of mass extinctions. *Scientific American,* July 1996, 72-78.

Espinasa, Luis & Espinasa, Monika. Why do cave fish lose their eyes? *Natural History,* June 2005, 44-49.

Evans, Scott; Tu, Chenyi; Rizzo, Adriana & Droser, Mary L. Environmental drivers of the first major animal extinction across the Ediacaran White Sea-Nama transition. PNAS, *119.* |no. 46. https://doi.org/10.1073/pnas.2207475119

Eysenck. Hans. *The Measurement of Intelligence.* Lancaster: MTP, 1973.

Fassett, James; Zielinski, Robert & Budahn, James. Dinosaurs that did not die: Evidence for Paleocene dinosaurs in Ojo alamo Sandstone, San Juan Basin, New Mexico. In Koeberl,

Christian and MacLeod (Ed) *Catastrophic Events and Mass Extinctions.* Boulder: Geological Society of America, 2002.

Fastovsky, David. Yes, and an asteroid did the deed. *Natural History,* May 2005, *114,* 52- 53.

Federoff, Nina & Botstein, David. *The Dynamic Genome.* Cold Springs: Colds Springs Laboratory Press, 1992)

Feibleman, James. Theory of integrative levels. *British Journal for the Philosophy of Science,* 1954, *5,* 59-66.

Feynman, Richard. "The relation of science and religion. In L. Jacobus (ed.) *A World of Ideas,* Boston: Bedford, 2002.

Field, Jeremy & Brace, Selina. Pre-social benefits of extended parental care. *Nature,* 8 April 2004, *428,* 650-652.

Font, Eric; Adatte, Thierry; Nobrega Sial, Alcides; Drude de Lacerda, Luiz; Keller, Gerta & Punekar, Jahnavi. Mercury anomaly, Deccan volcanism, and the end-Cretaceous mass Extinction. *Geology,* 2016, *44,* 171-174.

Forey, Peter. A home from home for coelecanths. *Nature,* September 24, 1998, *395,* 319- 320.

Fortey, Richard. *Trilobite!* New York: Alfred E. Knopf, 2000.

Foster, Jane & Neufeld, Karen-Anne. Gut-brain axis: How the microbiome influences anxiety and depression. *Trends in Neurosciences,* 2013, *36,* 305-312.

Fox-Genovese, Elizabeth. *Feminism is NOT the Story of My Life.* New York: Doubleday, 1996.

Fraga, M.; Ballestar, E.; Paz, M.; Ropero, S.; Setien, F.; Ballestar, M.; Heine-Suner, D.; Cigudosa, J.; Urioste, M.; Benitez, J.; Boix-Chomet, M.; Sanchez-Aguilera, A.; Ling, C.; Carlson, E.; Poulsen, P.; Vaag, A.; Stephen, Z.; Spector, T.; Wu, Z.; Plass, C. & Esteller, M. Epigenetic differences arise during the lifetime of monozygotic twins. *Proceedings of the Natural Academy of Sciences of the USA.* 2005, July 26, *102*, 10604-10609.

Francois, Roger. Cool stratification. *Nature,* 4 March 2004, 31-32.

Frederick, Michael & Gallup, Gordon. The demise of dinosaurs and learned taste aversions: The biotic revenge hypothesis. *Ideas in Ecology and Evolution*, July 7, 2018, *10,* 47-54.

French, Howard. E. O. Wilson's theory of everything. *The Atlantic Magazine*, October 2011.

Fricke, Hans & Hissmann, Karen. Natural habitat of coelecanths. *Nature,* July 26, 1990, *346,* 323-324.

Fukumori, Ichiro. NASA research offers explanation for Earth's bulging waistline. http://jpl.nasa.gov/releases/2002/release_2002_2 17.cfm/.

Furnes, Harald; Banerjee, Neil; Muehlenbachs, Karlis; Staudigel, Hubert; Maarten de Wit. Early life recorded in Archean pillow lavas. *Science,* 23 April 2004, *304,* 578-581.

Gagliano, Monica. On the spot: the absence of predators reveals eyespot plasticity in a marine fish. *Behavioral Ecology,* 2008, *19,* 733-739.

Gale, Barry. *Evolution Without Evidence.* Albuquerque: University of New Mexico Press, 1982.

Gallagher, William. Faunal changes across the Cretaceous-Tertiary (K-T) boundary in the Atlantic coastal plain of New Jersey: Restructuring the marine community after the K-T mass-extinction event. In Koeberl, Christian and MacLeod (Ed.) *Catastrophic Events and Mass Extinctions.* Boulder: Geological Society of America, 2002.

Gardner, A. & Gilmour, I. Organic geochemical investigation of terrestrial Cretaceous-Tertiary boundary successions from Brownie Butte, Montana, and the Raton Basin, New Mexico. In Koeberl, Christian and MacLeod (Ed.) *Catastrophic Events and Mass Extinctions.* Boulder: Geological Society of America, 2002.

Gardner, Eldon. *Principles of Genetics.* New York: John Wiley & Sons, 1975.

Garwin, Laura. In praise of interdisciplinarity. *Nature,* 17 August 1995, *376,* 547.

Geddes, Patrick. (1882) Further researches on animals containing chlorophyll. *Nature, 25,* 303-305.

Gerasimov, M. Toxins produced by meteorite impacts and their possible role in a biotic mass extinction. In Koeberl, Christian and MacLeod (Ed.) *Catastrophic Events and Mass Extinctions.* Boulder: Geological Society of America, 2002.

Gibbons, Ann. On the many origins of species. *Science,* 13 September 1996, *273,* 1496-1499.

---------The species problem. *Science,* 13 September 1996, *273,* 1501.

Gilbert, Scott; Sapp, Jan & Tauber, Alfred. A symbiotic view of life: we have never been individuals. *The Quarterly Review of Biology*, 2012, *87*, 325-341.

Gilliver, Moira; Bennett, Malcolm; Begon, Michael; Hazel, Sarah & Hart, C. Antibiotic resistance found in wild rodents. *Nature,* 16 September 1999, *401,* 233.

Gillman, Mark. *Envy as a Retarding Force in Science.* Aldershot: Avery, 1996.

Gladyshev, Eugene; Meselson, Matthew & Arkhipova, Irina. (2008) Massive horizontal gene transfer in bdelloid rotifers. *Science, 320*, 1210-3.

Goetz, Delia & Morley, Sylvanus. *Popol Vuh: The Sacred Book of the Ancient Quiche Maya.* Norman: University of Oklahoma Press, 1950.

Goldschmidt, Richard. (1940/1982) *The Material Basis of Evolution.* New Haven: Yale University Press.

Goldsmith, Donald. *Scientists Confront Velikovsky.* New York: W. W. Norton, 1977.

Gompert, Zacariah; Fordyce, James; Forister, Matthew; Shapiro, Arthur & Nice, Chris. Homoploid hybrid speciation in an extreme habitat. *Science,* 22 December 2006, *314,* 1923-1925.

Gould, Stephen Jay. Gould, Stephen Jay. Return of the hopeful monster. *Natural History*, 1977, *86,* 1-6.

---------*The Mismeasure of Man.* W. W. Norton, 1981.

---------*The Panda's Thumb.* New York: W.W. Norton, 1982.

--------- (1982) The uses of heresy. In Richard Goldschmidt's *The Material Basis of Evolution.* New Haven: Yale University Press.

---------*Bully for Brontosaurus.* New York: W. W. Norton, 1991.

---------& Eldredge, Niles. Punctuated equilibrium comes of age. *Nature,* 18 November 1993, *366,* 223-227.

---------*Dinosaur in a Haystack.* New York: Crown Trade Paperbacks, 1995.

---------*Wonderful Life.* New York: W. W. Norton, 2007.

---------*Punctuated Equilibrium.* Cambridge: Belknapp Press, 2007b.

Graham, Loren. *Lysenko's Ghost.* Cambridge: Harvard University Press, 2016.

Grant, Bob. (2019) Tides red in tooth and claw. *The Scientist, 33,* 27-33.

Gray, Peter. *The Irish Famine.* New York: Harry Abrams Publishers, 1995.

Green, Mel. Annals of mobile DNA elements in Drosophila: The impact and influence of Barbara McClintock. In Fedoroff, Nina & Botstein, David. *The Dynamic Genome.* Cold Spring: Cold Spring Harbor Laboratory Press, 1992.

Greene, John. *The Death of Adam.* New York: Mentor Books, 1961.

---------*Darwin and the Modern World View.* New York: Mentor Books, 1963.

Grimaldi, David & Engel, Michael. *The Evolution of Insects.* New York: University of Cambridge Press, 2005.

Guterl, Fred. Mission to Mercury. *Discover,* April 2004, *25,* 34-41.

Hackett, Robert. The Computer Maverick Who Modeled the Evolution of Life. *Nautilus,* July 11,

2019. http://nautil.us/issue/74/networks/the-computer-maverick-who-modeled-the-evolution-of-life

Hall, B. K. Epigenetics: regulation not replication. *Journal of Evolutionary Biology,* 1998, *11,* 201-205.

Hall, Stephen. James Watson and the search for biology's "Holy Grail." *Smithsonian,* February 1990, 41-49.

Harder, Ben. What caused Argentina's mystery craters? http://news.nationalgeographic.com/news/2002/05/0509_020509_glassmeteorite.html.

Harland, W. & Rudwick, Martin. The great Infra-Cambrian Ice Age. *Scientific American,* August 1964, 28-36.

Harlow, H. F. Mice, monkeys, men and motives. *Psychological Review,* 1953, *60,* 23-32.

---------Nature of law---simplified. *American Psychologist,* 1970, *25,* 161-168.

Heard, Edith & Martienssen, Robert. (2014) Transgenerational epigenetic inheritance: myths and mechanisms. *Cell, 157,* 95-109.

Heise, John. Personal communication, 2003.

Heller, Mikhail & Nekrich, Alexander. *Utopia in Power.* New York: Summit Books, 1986.

Hendry, Andrew; Wenburg, John; Bentzen, Paul; Volk, Erik & Quinn, Thomas. Rapid evolution of reproductive isolation in the wild: Evidence from introduced salmon. *Science,* 20 October 2000, *290,* 516-521.

Hildebrand, A; Pilkington, M.; Connors, M. Ortiz-Aleman, C. & Chavez, R. Size and structure of the Chicxulub crater revealed by horizontal gravity gradients and cenotes. *Nature,* 3 August 1995, *376,* 415-417.

Hiroyuki, Agawa. *The Reluctant Admiral.* Tokyo, Kodansha International, 1979.

Hodych, J. P. & Dunning, G. R. did the Manicouagan impact trigger end-of-Triassic mass extinction? *Geology,* 1992, *20,* 51-54.

Hoffer, Eric. *The True Believer.* New York: Perennial Library, 1951.

---------*The Passionate State of Mind.* New York: Perennial Library, 1955.

Hoffman, P. & Schrag, D. The snowball Earth hypothesis: testing the limits of global change. *Terra Nova,* 2002, *14,* 129-155.

Holland, Jennifer. Red alert. *National Geographic,* September 2007, *212,* 24.

Holloway, Ralph L. The Mismeasure of Science: Stephen Jay Gould versus Samuel George Morton on skulls and bias. *PLoS Biology*, June 2011, *9*, Issue 6, 1-6.

Huber, Brian; MacLeod, Kenneth & Norris, Richard. Abrupt extinction and subsequent reworking of Cretaceous planktonic foraminifera across the Cretaceous-Tertiary boundary: Evidence from the subtropical North America. In Koeberl, Christian and MacLeod (Ed.) *Catastrophic Events and Mass Extinctions*. Boulder: Geological Society of America, 2002.

Hull, David. A revolutionary philosopher of science. *Nature,* 18 July 1996, *382,* 203-204.

Hunt, Gene. The relative importance of directional change, random walks, and stasis in

the evolution of fossil lineages. *PNAS,* 20 February 2007, *104,* 18404-18408.
Huxley, Julian. Genetics: the real issue. *Nature,* June 25, 1949, *163,* 974-982.
----------*Evolution in Action.* New York: Signet Science Library Book, 1953.
Israde-Alcántara, Isabel; Bischoff, James; Domínguez-Vázquez; Li, Hong-Chun; DeCarli, Paul; Bunch, Ted; Wittke, James; Weaver, James; Firestone, Richard; West, Allen; Kennett, James; Mercer, Chris; Xie, Sujing; Richman, Eric; Kinzie, Charles & Wolbach, Wendy. Evidence from central Mexico supporting the Younger Dryas extraterrestrial impact hypothesis. *PNAS*, 5 March 2012. http://www.pnas.org/content/109/13/E738.full
Jablonaka, Eva & Lamb, Marion. "Transgenerational epigenetic inheritance." In Pigliucci, Massimo & Müller, Gerd. (Eds.) *Evolution: The Extended Synthesis.* Cambridge: The MIT Press, 2010.
Jablonski, David. Origination patterns and multilevel processes in macroevolution. In Pigliucci, Massimo & Müller, Gerd. (Eds.) *Evolution: The Extended Synthesis.* Cambridge: The MIT Press, 2010.
Jacobsen, S. E. & Meyerowitz, E. M. Hypermethylated SUPERMAN epigenetic alleles in *Arabidopsis. Science,* 1997, *277,* 1100-1103.
Jaffe, Mark. *The Gilded Dinosaur.* New York: Crown Publishers, 2000.
James, Carollyn. *Science 81,* May 1981, 66.
Jastrow, Robert. *Red Giants and White Dwarfs.*

Evanston, Harper & Row, 1967.

Jeske, J. & Whitten, M. Motivational Distortion of the Sixteen Personality Factor questionnaire by persons in job applicants' roles. *Psychological Reports*, 1975, *37,* 1192-1193.

Johnsen, Sigfŭs; Clausen, Henrik; Dansgaard, Willi; Gundestrup, Niels; Hammer, Claus; Andersen, Uffe; Andersen, Katrine; Hvidberg, Christine; Dahl-Jensen, Dorthe; Steffensen, Jørgen; Shoji, Hitoshi; Sveinbjöornsdóttir, Ărni; White, Jim; Jouzel, Jean & Fisher, David. The δ^{18}O record along the Greenland Ice Core Project deep ice core and the problem of possible Eemian climatic instability. *Journal of Geophysical Research,* 1997, *102*, 26,397-26,410.

Johnson, Kirk & Murphy, Dean. Drought-plagued West is awash in worry. *San Antonio Express-News* 2 May 2004, 5A.

Jones, Brian. William Herschel: Pioneer of the stars. *Astronomy,* November 1988, *54,* 40- 53.

Kaiser, Joyce. NASA ducks search for celestial threats. *Science,* 18 August 1995, *269,* 911.

Kale, Vivek & Pande, Kanchan. (2022) Reappraisal of duration and eruptive rates in Deccan Volcanic province, India. *Jour. Geol. Soc. India, 98*, 7-17.

Kamber, Balz. First evidence for early meteorite bombardment of Earth. http://news.nationalgeographic.com/news/2002/07/0725_020725_meteor.html.

Kanipe, Jeff; Talcott, Richard & Burnham, Robert. The rise and fall of the sun's activity. *Astronomy,* October 1988, 22-31.

Keller, Evelyn. *A Feeling for the Organism*. New York: W. H. Freeman and Company, 1983.

Keller, Gerta; Adatte, Thierry; Stinnesbeck, Wolfgang; Affolter, Mark; Schilli, Lionel & Lopez-Olivia, José. Multiple layers in the late Maastrichtian of northeast Mexico. In Koeberl, Christian and MacLeod (Ed.) *Catastrophic Events and Mass Extinctions*. Boulder: Geological Society of America, 2002.

----------Punekar, Jahnavi & Mateo, Paula. (2015) Upheavals during the Late Maastrichtian: Volcanism, climate and faunal events preceding the end-Cretaceous mass extinction. *Palaeogeography, Palaeoclimatology, Palaeoecology, 441*, 137-151.

--------- Mateo, Paula; Monkenbusch, Johannes; Thibault, Nicolas; Punekar, Jahnavi; Spangenberge, Jorge; Abramovich, Sigal; Ashckenazi-Polivoda, Sarit; Schoenea, Blair; Eddy,Michael; Sampertoni; Kyle; Khadri, Syed & Adatte, Thierry. (2020) ercury linked to Deccan Traps volcanism, climate change and the end Cretaceous mass extinction. *Global and Planetary Change*,194, 103312. https://doi.org/10.1016/j.gloplacha.2020.103312

Kelley, Patricia. Evolutionary patterns of eight Chesapeake Group mollusks: Evidence for the model of punctuated equilibria. *Journal of Paleontology*, 1983, *57*, 581-598.

Kemp, Martin. Shelley's shocks. *Nature*. August 1998, *394*, 529.

Kennett, D.; Kennett, J.; West, A.; Mercer, C.; Hee, S.;

Bement, L.; Bunch, T.; Sellers, M. & Wolbach, W. Nanodiamonds in the Younger Dryas Boundary sediment layer. *Science,* 2 January 2009, *323,* 94.

Kerney, Ryan; Kim, Eunsoo; Hangarter, Roger; Heiss, Aaron; Bishop, Cory & Hall, Brian. (2011) Intracellular invasion of green algae in a salamander host. *Proceedings of the National Academy of Sciences of the USA, 10,* 6497-6502.

Kerr, Richard. New mammal data challenge evolutionary pulse theory. *Science,* 26 July 1996, *273,* 431-432.

---------Chesapeake Bay impact crater confirmed. *Nature,* 22 September 1995 (a), *269,* 1672.

---------Animal oddballs brought into the ancestral fold? *Science,* 27 October 1995 (b), *270,* 580-581.

---------Did volcanoes drive ancient extinctions? *Science,* 18 August 2000, *289,* 1130-1131.

---------Mega-eruptions drove the mother of mass extinctions. *Science,* 20 December 2013, *342,* 1424.

---------Cosmic dust supports a Snowball Earth. *Science,* 8 April 2005, 181.

Khakhina, Liya. *Concepts of Symbiogenesis.* New Haven: Yale University Press, 1992.

Kideys, Ahmet. Fall and rise of the Black Sea ecosystem. *Science,* 30 August 2002, 1482-1484.

Kjaer, Kurt; Larsen, Nicolaj; Binder, Tobias; Bjørk, Anders; Eisen, Olaf; Fahnestock, Mark &Funder, Sven. A large impact crater beneath Hiawatha Glacier in northwest Greenland.

Science Advances, 14 November 2018, *4.* DOI: 10.1126/sciadv.aar8173

Knoll, Andrew. *Life on a Young Planet.* Princeton: Princeton University Press, 2003.

Köberl, Christian. The geochemistry and cosmochemistry of impacts. In *Treatise of Geochemistry.* 2007, 1-52. DOI: 10.1016/B978-008043751-4/00228-5

----------Farley, K.; Peucker-Ehrenbrink, B. & Sephton, Mark. Geochemistry of the end-Permian extinction event in Austria and Italy: No evidence for an extraterrestrial component. *Geology,* 2004, *32,* 1053-1056.

----------Poag, C.; Reimold, Wolf & Brandt, Dion. Impact origin of the Chesapeake Bay structure and the source of the North American tektites. *Science,* 1 March 1996, *271,* 1263-1266.

Koestler, Arthur. *The Case of the Midwife Toad.* New York: Random House, 1971.

----------*The God that Failed.* New York: Bantam, 1965.

Krajick, Kevin. Discoveries in the dark. *National Geographic,* September 2007, *212,* 134-147.

Kramer, Peter & Bressan, Paola. (2015) Humans as superorganisms: how microbes, imprinted genes, and other selfish entities shape our behavior. *Perspectives on Psychological Science 10,* 464-481.

Kring, David & Durda, Daniel. The day the world burned. *Scientific American,* December 2003, *289,* 98-105.

Kruuk, Hans. *Niko's Nature.* Toronto: Oxford University

Press, 2003.

Kuhn, Thomas. Historical structure of scientific discovery. *Science,* 1 June 1962, *136,* 760-764.

Kwon, Diana. Effects of antidepressants span three generations in fish. *The Scientist,* December 10,2018. https://www.the-scientist.com/news-opinion/effects-of-antidepressants-span-three-generations-in-fish-65193

Kyte, Frank. A meteorite from the Cretaceous/Tertiary boundary. *Nature,* 19 November 1998, *396,* 237-239.

Leach, Mark & Givnish, Thomas. Ecological determinants of species loss in remnant prairies. *Science,* 13 September 1996, *273,* 1555-1558.

Lessem, Don. *Kings of Creation.* New York: Simon & Schuster, 1992.

Levine, Michael; Cattoglio, Claudia & Tijan, Robert. (2014) Looping back to leap forward: transcription enters a new era. *Cell, 157,* 13-25.

Levy, David. *Impact Jupiter.* New York: Plenum Press, 1995.

Lewis, J ason E.; DeGusta, David; Meyer, Marc R.; Monge, Janet M.; Mann, Alan E. & Ley, Willy. *Ranger to the Moon.* New York: Signet Science Library, 1965.

Liddell, Studio. Impact! *Maxim,* April 2004, *8,* 62-63. Lira-Medeiros, Caterina; Parisod, Christian; Fernandes, Avancini; Mata, Camila; Cardoso, Monica & Ferreira, Paulo. Epigenetic variation in mangrove plants occurring in contrasting natural environment. *PLoS ONE,* 2010, *5,* 1-8.

Lloyd, Francis. *The Carnivorous Plants.* New York: The

Ronald Press Company, 1942.
Loder, Natasha. Wallace rescued from a grave injustice. *Nature,* August 5, 1999, *400,* 489.
Lorenz, Konrad. *The Natural Science of the Human Species.* Cambridge: MIT Press, 1996.
Losos, Jonathan & Schluter, Dolph. Analysis of an evolutionary species-area relationship. *Nature,* 2000, *408,* 847-850.
----------Schoener, Thomas; Langerhans, Brian & Spiller, David. Rapid temporal reversal in predator-driven Natural Selection. *Science,* 17 November 2006, *314,* 111.
Lovelock, James. *The Ages of Gaia.* New York City: W. W. Norton, 1988.
Lyell, Charles. *Principles of Geology, Being an Attempt to Explain the Former Changes of the Earth's Surface by Reference to Causes Now in Operation.* London: Penguin Books, 1833/1997.
Lyte, Mark. (2013) Microbioal endocrinology in the microbiome-gut-brain axis: How bacterial production and utilization of neurochemicals influence behavior. *PLOS, 9,* Issue 11.
MacArthur, Robert & Wilson, Edward. *The Theory of Island Biogeography.* Princeton: Princeton University Press, 1967/2001.
Machan, Tibor. *The Pseudoscience of B. F. Skinner.* New Rochelle: Arlington House, 1974.
Maeda, Taro; Hirose, Euichi; Chikaraishi, Yoshito; Kawato, Masaru; Takishita, Kiyotaka; Yoshida, Takao; Verbruggen, Heroen; Tanaka, Jiro; Shimamura, Shegeru; Takaki, Yoshihiro; Tsuchiya, Masashi; Iwai, Kenji & Maruyama,

Tadashi. (2012) Algivore or phototroph? Plakobranchus ocellatus (Gastropoda) continuously acquires kleptoplasts and nutrition from multiple algal species in nature. *PLOS one,* 7, 1-12.

Margulis, Lynn. Symbiosis and evolution. *Scientific American,* 1971, *225,* 48-61.

-----------*Symbiotic Planet.* New York: Basic Books, 1998.

-----------& Sagan, Dorion. *Microcosmos.* Berkeley: University of California Press, 1997.

----------How I became a scientist. *Natural History,* 2004, *113,* 80.

----------& Sagan, Dorion. *Acquiring genomes.* New York: Basic Books, 2002.

Marler, Peter & Hamilton, William. *Mechanisms of Animal Behavior.* New York: John Wiley & Sons, 1966.

Marks, Richard. *Three Men of the Beagle.* New York: Avon Books, 1991.

Matsui, T.; Imamura, F.; Tajika, E.; Nakano, Y. & Fujisawa, Y. Generation and propagation of a tsunami from the Cretaceous-Tertiary Impact Event. In Koeberl, Christian and MacLeod (Ed.) *Catastrophic Events and Mass Extinctions.* Boulder: Geological Society of America, 2002.

Mayell, Hillary. Did plants cool the Earth and spark explosion of life? http://news.nationalgeographic.com/news/2001/08/0810_preciousplants.html.

Mayor, Adrienne. Tales from the Badlands. *Natural History,* May 2005, *114,* 56-61.

Mayr, Ernst. Speciational evolution of punctuated equilibria. http://www.sphenjaygould.org/library/mayr_punctuated.html.

--------- *What Evolution Is.* New York: Perseus Books, 2001.

Mehos, Donna. Ivan E. Wallin and his theory of symbionticism. In Lynn Margulis & Mark McMenamin (Eds.) *Concepts of Symbiogenesis.* New Haven: Yale University Press, 1992. pp. 149-163.

McBride, Chris. *The White Lions of Timbavati.* London: Paddington Press, 1977.

McCarthy, Deirdre; Morgan, Thomas; Lowe, Sarah; Williamson, Matthew; Spencer, Thomas; Biederman, Joseph & Bhide, Pradeep. Nicotine exposure of male mice produces behavioral impairment in multiple generations of descendants. *PLOS Biology,* October 16, 2018. https://journals.plos.org/plosbiology/article?id=10.1371/journal.pbio.2006497

McClintock, Barbara. The origin and behavior of mutable loci in maize. *Proceedings of the National Academy of Science,* 1950, *36,* 344-355.

----------- Some parallels between gene control systems in maize and in bacteria. *The American Naturalist,* September 1961, *95,* 265-277.

----------- The contribution of one component of a control system to versatility of gene expression. *Carnegie Institution of Washington Yearbook.* 1971, *70,* 5-17.

----------- Mechanisms that rapidly reorganize the genome. *10th Stadler Genetic Symposium,* 1978,

25-48.

---------- Modified gene expressions induced by transposable elements. In W. Scott, et. al., (Ed) *Mobilization and Reassembly of Genetic Information.* New York: Academic Press, 1980, pp. 11-19.

---------- The significance of responses of the genome to challenge. *Science,* 16 November 1984, *226,* 792-801.

McKinney, Michael. Understanding evolution: The next step. *Science*, 6 September 1996, *273,* 1347.

Medvedev, Roy. *Let History Judge.* New York: Alfred Knopf, 1971.

Medvedev, Zhores. *The Rise and Fall of T. D. Lysenko.* Garden City: Doubleday & Co. 1971.

---------- & Medvedev, Roy. *A Question of Madness.* New York: Vintage Books, 1971.

Melosh, H. Around and around we go. *Nature,* 3 August 1995, 386-387

Meyer, Axel. Learning from the Altmeister. *Nature,* 29 April 2004, *428,* 897.

Miele, Frank. The Ionian instauration. *Skeptic,* 1998, *6,* 76-85.

Mika, Aggie. Government nixes teaching evolution in Turkish schools. *The Scientist,* June 23, 2017, http://www.the-scientist.com/?articles.view/articleNo/49732/title/Government-Nixes-Teaching-Evolution-in-Turkish-Schools/&utm_campaign=NEWSLETTER_TS_The-Scientist-Daily_2016&utm_source=hs_email&utm_mediu

m=email&utm_content= 53553238&_hsenc=p2ANqtz-8LJZVG52

Miljković, Katarina; Collins, Gareth; Mannick, Sahil & Bland, Philip. Morphology and population of binary asteroid impact craters. *Earth and Planetary Science Letters,* February 1, 2013, *363,* 121-132.

Miller, Walter. *A Canticle for Leibowitz.* Philadelphia: J. B. Lippincott Company, 1959.

Mizrahi, Itzhak & Kokou, Fotini. More than the sum of its parts. *The Scientist,* 2019, *33,* 21.

Mlot, Christine. Microbes hint at a mechanism behind punctuated evolution. *Science,* 21 June 1996, *272,* 1741.

Moore, Peter. Woodpecker population drills. *Nature,* 10 June 1999, *399,* 528-529.

Moore, Ruth. *Evolution.* New York: Time-Life Books, 1964.

Morell, Virginia. Amazonian diversity: A river doesn't run through it. *Science,* 13 September 1996, *273,* 1496-1497.

---------Genes vs. teams: Weighing group tactics in evolution. *Science,* 9 August 1996, *273,* 739-740.

---------Starting species with third parties and sex wars. *Science,* 13 September 1996, *273,* 199-1502.

Moritz, Richard; Bitz, Cecilia & Steig, Eric. Dynamics of recent climate change in the Arctic. *Science,* 30 August 2002, *297,* 1497-1502.

Morris, Desmond. *The Naked Ape.* New York: Dell, 1967.

Morris, Richard. *The Evolutionists.* New York: W. H. Freeman, 2001.

Moser, Don. *Central American Jungles.* Chicago: Time-Life Books, 1975.

Mulholland, J. Derral. The Chandler Wobble. Smithsonian? 134-141.

Müller, Gerd. Epigenetic innovation. In Pigliucci, Massimo & Müller, Gerd (Eds.) *Evolution: The Extended Synthesis.* Cambridge: The MIT Press, 2010.

Müller, Gerd & Newman, Stuart. The innovation triad: An EvoDevo agenda. *Journal of Experimental Zoology,* 2005, *304 B, 487-503*.

Nasar, Sylvia. *A Beautiful Mind.* New York: Simon& Schuster, 1998.

Nass, M. M. (1969) Uptake of isolated chloroplasts by mammalian cells. *Science, 165,* 1128-1131.

Nätt, Daniel; Barchiesi, Riccardo; Murad, Josef; Jian Feng; Nestler, Eric J.; Champagne, Frances A. & Annika Thorsell. Perinatal Malnutrition Leads to Sexually Dimorphic Behavioral Responses with Associated Epigenetic Changes in the Mouse Brain. *Scientific Reports, 2017, 7,* Article number: 11082

Nature. 100 years ago. *Nature,* 22 October 1998, *395,* 751.

Neilson, Susie. Is Japanese culture traumatized by centuries of natural disaster? *Nautilus,* August 30, 2017, http://nautil.us//blog/-is-japanese-culture-traumatized-by-centuries-of-natural-disaster?utm_source=Nautilus&utm_campaign=5855f4dded-EMAIL_CAMPAIGN_2017_11_10&utm_medium=email&utm_term=0_dc96ec7a9d-

Nelson, O. & Klein, A. Characterization of an *spm*-controlled bronze-mutable allele in maize. *Genetics,* 1984, *106,* 769-779.

Nicolle, Jacques. *Louis Pasteur.* New York: Fawcett Premier Books, 1961.

Nimmo, F.; Hart, S.; Korycansky, D. & Agnor, C. Implications of an impact origin for the Martian hemispheric dichotomy. *Nature,* 2008, *453,* 1220-1224.

Norris, R. & Firth, J. Mass wasting of Atlantic continental margins following the Chicxulub impact event. In Koeberl, Christian and MacLeod (Ed.) *Catastrophic Events and Mass Extinctions.* Boulder: Geological Society of America, 2002.

Novikoff, Alex. The concept of integrative levels and biology. *Science,* 2 March 1945, *101,* 209-215.

Odajnyk, Volodymyr. *Jung and Politics.* New York: Harper Colophon, 1976.

Olsen, Paul. Giant lava flows, mass extinctions, and mantel plumes. *Science,* 23 April 1999, *284,* 604-605.

Olsen, P.E.; Shubin, N.H. & Anders, M.H. New Early Jurassic tetrapod assemblages constrain Triassic-Jurassic tetrapod extinction event. *Science,* 28 August 1987, *237,* 1025-1029.

----------Kent, D.V.; Sues, H.D.; Köberl, C; Huber, H.; Montanari, A.; Rainforth, E.C. Fowell, S.J.; Szajma, M.J. & Hartline, B.W. Ascent of dinosaurs linked to an iridium anomaly at the Triassic-Jurassic boundary. *Science,* 17 May 2002, *296,* 1305-1307.

O'Neill, Scott; Hoffmann, Ary & Werren, John. (Eds.) *Influential Passengers.* Oxford: Oxford University Press.

Oppenheimer, Stephen. *Eden in the East.* London: Wedenfeld & Nicolson, 1998.

Orlando, Ludovico & Willerslev, Eske. An epigenetic window into the past? *Science,* 2014, *345,* 511.

Orr, H. A. The descent of Gould. *The New Yorker,* September 30, 2002.

Orwell, George. *Coming Up for Air.* San Diego: Harcourt Brace Jovanovich, 1939.

Packard, Alphaeus. *Lamarck the Founder of Evolution.* London: Longmans, Green & Co. 1901.

Pagel, Mark; Venditti, Chris &Meade, Andrew. Large punctuational contribution on speciation to evolutionary divergence at the molecular level. *Science,* 6 October 2006, *314,* 119-121.

Paglia, Camille. *Vamps & Tramps.* New York: Random House, 1994.

Palmer, Douglas. Ediacarans in deep water. *Nature,* 11 January 1996, *379,* 114.

Palumbi, Stephen. *The Evolution Explosion.* New York: W. W. Norton, 2001.

Parthasarathy, G.; Bhandari, N.; Vairamani, M.; Kunwar, A.7 Narasaiah, B. In Koeberl, Christian and MacLeod (Ed) *Catastrophic Events and Mass Extinctions.* Boulder: Geological Society of America, 2002.

Parker, Andrew. *In the Blink of an Eye.* Cambridge: Perseus Publishing, 2003.

Parsell, D. Mass extinction that led to age of dinosaurs

was swift, study shows. http://news.nationalgeographic.com/news/2001/05/0510_massex.html.

Pasteur, Louis. The germ theory and its applications to medicine and surgery. In Pasteur, L. & Lister, J. *Germ Theory and its Applications to Medicine & On the Antiseptic Principle of the Practice of Surgery.* Amherst: Prometheus Books Great Minds Series, 1996.

---------The physiological theory of fermentation. In Pasteur, L. & Lister, J. *Germ Theory and its Applications to Medicine & On the Antiseptic Principle of the Practice of Surgery.* Amherst: Prometheus Books Great Minds Series, 1996.

Patterson, D. J. *Free-Living Freshwater Protozoa.* Toronto: John Wiley &Sons, 1996.

Pearl, Mary. Ecologists in Sri Lanka assess the impact of the 2004 tsunami. *Discover,* October 2006, *27,* 29.

Pearson, Helen. What is a gene? *Nature,* 2006, *441,* 399-401.

Pearson, Patricia. *When She Was Bad.* New York: Viking, 1997.

Pellegrino, Charles. *Unearthing Atlantis.* New York: Vintage Books, 1993.

Penn, J.; Deutsch, C.; Payne, J. & Sperling, Erik A. Temperature-dependent hypoxia explains biogeography and severity of end-Permian marine mass extinction. *Science,* 2018, *362,* 1327-1329.

Pennisi, Elizabeth. Pennisi, Elizabeth. How beach life favors blond mice. *Science,* 2009, *329,* 1330-

1333.

----------Cavefish supports controversial evolutionary mechanism. *Science,* 13 December 2013, *342,* 1304.

----------Disputed islands. *Science,* 8 August, 2014, *345,* 611-613.

----------Evolution heresy? Epigenetics underlies heritable plant traits. *Science,* 2010, *341,* 1055.

Perlman, Dan & Adelson, Glenn. *Biodiversity.* Cambridge: Blackwell Science, 1997.

Perkins, Sid. A century later, scientists still study Tunguska. *Science News,* 2008, *10,* 5-6.

Pierce, Sidney; Biron, Rachel & Rumpho, Mary. (1996) Endosymbiotic chloroplasts in molluscan cells contain proteins synthesized after plastic capture. *The Journal of Experimental Biology, 199,* 2323–2330.

Pinter, Nicholas & Ishman, Scott. Impacts, mega-tsunami, and other extraordinary claims. *GSA Today,* 2008, *18,* 37.

Pliny, Younger. *The Letters of the Younger Pliny.* Baltimore: Penguin Books, 113/1969.

Preisinger, Anton; Asianian, Selma; Brandstätter, Franz; Grass. Friedrich; Stradner, Herbert & Summesberger, Herbert. Cretaceous-Tertiary profile, rhythmic deposition, and geomagnetic polarity reversals of marine sediments near Bjala, Bulgaria. In Koeberl, Christian and MacLeod (Ed.) *Catastrophic Events and Mass Extinctions.* Boulder: Geological Society of America, 2002.

Pringle, Peter. *The Murder of Nikolai Vavilov.* Toronto: Simon & Schuster, 2008.

Prothero, Donald. Punctuated equilibrium at twenty: A paleontological perspective. http:// www.skeptic.com/01.3.prothero-punc-eq.html.

Quammen, David. *The Song of the Dodo.* New York: Scribner, 1996.

-------------- Was Darwin wrong? *National Geographic,* November 2004, 4-35.

Ramón y Cajal, Santiago. *Advice for a Young Investigator.* Cambridge: Bradford Book, 1887/1999.

Ramos, Pedro. *The Cave of Altamira.* New York: Harry Abrams, Inc., 1998.

Rassoulzadegan, Minoo; Grandjean, Valérie; Gounon, Pierre; Vincent, Stéphane; Gillot, Isabelle & Cuzin, Francois. RBA-mediated non-mendelian inheritance of an epigenetic change in the mouse. *Nature,* 2006, *441,* 469-474.

Raukas Anto. Postglacial impact events in Estonia and their influence on people and their environment. In Koeberl, Christian and MacLeod (Ed.) *Catastrophic Events and Mass Extinctions.* Boulder: Geological Society of America, 2002.

Raup, David. *The Nemesis Affair.* New York: W. W. Norton, 1999.

--------- *Extinction, Bad Genes or Bad Luck?* New York: W. W. Norton, 1992.

Raven, Peter. A multiple origin for plastids and mitochondria. *Science,* 1970, *169*, 641-646.

Reebs, Stephan. Evolutionary circles. *Natural History,* April 2004, *113,* 12.

Refrew, Colin. Kings, tree rings and the Old World. *Nature,* 27 June 1996, *381,* 733-734.

Reichhardt, Tony. Asteroid watchers debate false alarm. *Nature,* 19 March 1998, *392,* 215.

Reif, Wolf-Ernst. The search for a macroevolutionary theory in German paleontology. *Journal of the History of Biology,* 1986, *19,* 79-130.

Reimold, Wolf Uwe. The Impact Crater Bandwagon. *Meteoritics & Planetry Science,* 2007, *42,* 1467-1471.

Reinheimer, Hermann. (1915/2012) *Symbiogenesis: The universal law of progressive evolution.* Kingston-on-Thames: Forgotten Books.

Renne, Paul; Zhang, Zichao; Richards, Mark; Black, Michael & Basu, Asish. Synchrony and causal relations between Permian-Triassic Boundary crises and Siberian flood volcanism. *Nature,* 8 September 1995, 1413-1416.

Rensberger, Boyce. Death of dinosaurs: The true story? *Science Digest,* May 1986, *94,* 28-78.

Rhine, Louisa. *Psi---What is it?* San Francisco: Perennial Library, 1975.

Rhoades, Marcus. The early years of maize genetics. In Fedoroff, Nina & Botstein, David (Ed.) *The Dynamic Genome.* Cold Spring: Cold Springs Harbor Laboratory Press, 1992.

Richards, Eric. Inherited epigenetic variation---revisiting soft inheritance. www.hsph.Harvard.edu/niehs/metals Richards_Nature_2006.pdf

Richards, Robert. *The Meaning of Evolution.* Chicago: University of Chicago Press, 1992.

Richardson, Aaron & Palmer, Jeffrey. Horizontal gene transfer in plants. *Journal of Experimental*

Botany, 2006, *58,* 1-9.

Roach, John. Fossil challenge theory of rapid animal evolution in Cambrian. http://news.nationalgeographic.com/news/2001/07/0719_crustacean.html.,2001.

---------Killer asteroids: A real but remote risk? http://news.nationalgeographic.com/news/2003/06/0619_030619_killerasteroids.html.

---------Fossil leaves suggest asteroid killed dinosaurs. http://news.nationalgeographic.com/news/2002/06/0617_020617_fossilleaves.html.

Robbins, Michael W. Global warming triggers genetic change in red squirrels. *Discover,* January 2004, *25,* 35.

Robinson, Gene n& Barron, Andrew. Epigenetics and the evolution of instincts. *Science, 356,* 26-27.

Rohner, Nicolas; Jarosz, Dan F.; Kowalko, Johanna E.; Yoshizawa, Masato; Jeffery, William R.; Borowsky, Richard L.; Lindquist, Susan & Tabin, Clifford J. Cryptic variation in morphological evolution: HSP90 as a capacitor for loss of eyes in cavefish. *Science,* 13 December 2013, *342,* 1372-1375.

Romerstein, Herbert & Breindel, Eric. *The Venona Secrets.* Washington D.C: Regnery Publishing, 2000.

Rona, Peter. Secret survivor. *Natural History,* September 2004, *113,* 50-55.

Rose, Lynn. The censorship of Velikovsky's interdisciplinary synthesis. In Talbott, Stephen (Ed.) *Velikovsky Reconsidered.* New York: Warner Books, 1976.

Rovin, Jeff. *Science Fiction Films.* Secaucus: Citadel Press, 1975.

Rubinsky, Yuri & Wiseman, Ian. *A History of THE END of the World.* New York: Quill, 1982.

Rudwick, Martin. *Georges Cuvier, Fossil Bones, and Geological Catastrophes.* University of Chicago Press, 1997.

Rugg, Gordon & D'Agnese. *Blind Spot.* New York: Harper One, 2013.

Rutherford, Suzanne & Lindquist, Susan. Hsp90 as a capacitor for morphological evolution. *Nature,* 26 November 1998, *396,* 336-342.

Sabbagh, Karl. *A Rum Affair.* New York: Da Capo Press, 1999.

Saether, Bernt-Erik. Top dogs maintain diversity. *Nature,* 5 August 1999, *400,* 510-511.

Sagan, Carl. *The Dragons of Eden.* New York: Ballantine Books, 1977(a).

----------An Analysis of *Worlds in Collision.* In Donald Goldsmith (Ed.) *Scientists Confront Velikovsky.* New York: W. W. Norton, 1977(b).

Sagan, Dorion, (Ed.) *Lynn Margulis: The Life and Legacy of a Scientific Rebel.* White River Junction: Chelsea Green, 2012.

Sagan, Lynn. On the origin of mitosing cells. *J. Theoretical Biology,* 1967, *14,* 225-274.

Saleh, Dennis. *Science Fiction Gold.* Toronto: McGraw-Hill Book Company, 1979.

Santos, Levin. Downsizing evolution. *The Sciences,* March/April 1997, 46.

Self, Stephen; Mittal, Tushal; Dole, Gauri & Vanderkluysen, Loÿc. (2022) Toward understanding Deccan volcanism. *Annual Review of Earth and Planetary Sciences, 50,* 477–506.

Schaechter, Moselio. Lynn Margulis (1938-2011). *Science,* 2012, *335,* 302.

Shapiro, James (1992) Natural genetic engineering in evolution. *Genetica, 86,* 99-111.

-----------(1999) Transposable elements as the key to a 21st century view of evolution. *Genetica, 107,* 171-179.

-----------(2005) A 21st century view of evolution: genome system architecture, repetitive DNA, and natural genetic engineering. *Gene, 345,* 91-100.

Schiermeier, Quirin. Greenland's climate: A rising tide. *Nature,* 11 March 2004, *428,* 114-115.

Schilthuizen, Menno. "A paradox to all but himself." *Natural History,* September 2004, *113,* 58-62.

Schmitz, Birger; Tassinari, Mario & Peucker-Ehrenbrink, Berhnard. A rain of ordinary chondritic meteorites in the early Ordovician. *Earth and Planetary Science Letters,* 2001, *194,* 1-15.

-----------Harper, David A. T.; Peucker-Ehrenbrink, Bernhard; Stouge, Svend; Alwmark, Carl; Cronholm, Anders; Bergström, Stig M.; Tassinari, Mario & Wang, Xiaofeng. Asteroid breakup linked to the Great Ordovician Biodiversification Event. *Nature Geoscience,* 2008, *1,* 49–53.

Schindewolf, Otto. *Basic Questions in Paleontology.* Chicago: University of Chicago Press, 1950/1994.

Schmidt. Karsten. Lamarckism revisited. May 2007.

www.sequenom.com/Overview Epigenetics/OverviewEpigenetics.pdf

Schnell, Donald. *Carnivorous Plants of the United States and Canada.* Winston-Salem: John F. Blair, 1976.

Shoemaker, E. & Chao, E. New evidence for the impact origin of the Ries Basis, Bavaria, Germany. *Journal of Geophysical Research,* 1961, *66,* 3371-3378.

Schoener, Thomas & Spiller, David. Devastation of prey diversity by experimentally introduced predators in the field. *Nature,* 20 June 1996, *381,* 691-694.

Schuraytz, Benjamin; Lindstrom, David; Marin, Ruis; Martinez, Rene; Mittlefehldt, David; Sharpton, Virgil & Wentworth, Susan. Iridium metal in Chicxulub impact melt: Forensic chemistry on the K-T smoking gun. *Science,* 15 March 1996, *271,* 1573-1576.

Schwartz, Jeffrey. *Sudden Origins.* Singapore: John Wiley & Sons, 1999.

Schwartz, Randall. *Carnivorous Plants.* New York: Avon, 1974.

Secord, James. Introduction. In Lyell, Charles. *Principles of Geology.* London: Penguin Classics, 1997.

Shelley, Mary. *Frankenstein.* New York: Bantam 1818/1991.

Shermer, Michael. *In Darwin's Shadow.* Singapore: Oxford University Press, 2002.

Sigerist, Henry. *The Great Doctors.* New York: Dover, 1971.

Signor, Philip & Lipps, Jere. Sampling bias, gradual extinction patterns and catastrophes in the fossil record. *Geological Society of America*, 1982, Special Paper 190, 291-296.

Sikorski, Radek. *Dust of the Saints.* New York: Paragon House, 1990.

Simón, Armando. Sensitivity of the 16PF Motivational Distortion scale to response bias. *Psychological Reports,* 2007, *101,* 482-484.

Simonson, B.; Byerly, G. & Lowe, D. The early PreCambrian stratigraphic record of large extraterrestrial impacts. In (Ed.) Eriksson, P. G.; Altermann, Wladyslaw; Nelson, D. R.; Mueller, W. U.; Catuneanu, O. & Catuneanu, Octavian. *The Precambrian Earth: Tempos and Events (Developments in Precambrian Geology, Volume 12).* London: Elsevier Science, 2004.

Singer, S. F. How did Venus lose its angular momentum? *Science,* 11 December 1970, *170,* 1196-1198.

Sill, William. Fast track evolution in stressed environments the key factor for the origin of dinosaurs and mammals. http://www.earthwatch.org/pubaffairs/news/sill_tech.html

Skála, Roman; Ederová, Jana; Matêjka, Pavel & Hörz, Friedrich. In Koeberl, Christian and MacLeod (Ed.) *Catastrophic Events and Mass Extinctions.* Boulder: Geological Society of America, 2002.

Skinner, B. F. *About Behaviorism.* New York: Alfred Knopf, 1974

---------- *Beyond Freedom and Dignity.* New York:

Bantam, 1971.

Skinner, Michael. Unified theory of evolution. *Aeon*, 2017, https://aeon.co/essays/on-epigenetics-we-need-both-darwin-s-and-lamarck-s-theories

Smith, Charles. *Alfred Russel Wallace: A Capsule Biography*.http://www.wku.edu/~smithch/wallace/BIO.htm.

Sommers, Christina. *Who Stole Feminism?* Singapore: Simon & Schuster, 1994.

Somvanshi, Vishal; Sloup, Rudolph; Crawford, Jason; Martin, Alexander; Heidt, Anthony; Kim Kwi-suk; Clardy, Jon & Ciche, Todd. Single promoter inversion switches photorhabdus between pathogenic and mutualistic states. *Science, 337,* 2012, 88-93.

Soria-Carrasco, Victor; Gompert, Zachariah; Comeault, Aaron; Farkas, Timothy; Parchman, Thomas; Johnston, J. Spencer; Buerkle, C. Alex; Feder, Jeffrey; Bast, Jens; Schwander, Tanja; Egan, Scott; Crespi, Bernard & Nosil, Patrik. Stick insect genomes reveal natural selection's role in parallel speciation. *Science*, 16 May 2014, *344*, 738-742.

Soule, Gardner. *The Maybe Monsters.* New York: G. P. Putnam's Sons, 1963.

Southwood, Richard. *The Story of Life.* Oxford: Oxford University Press, 2003.

Spiller, David & Schoener, Thomas. A terrestrial field experiment showing the impact of eliminating top predators on foliage damage. *Nature,* 4 October 1990, *347,* 469-472.

Spray, John; Kelley, Simon & Rowley, David. Evidence

for a late Triassic multiple impact event on Earth. *Nature,* 12 March 1998, *392,* 171-173.

Srb, Adrian; Owen, Ray & Edgar, Robert. *General Genetics.* San Francisco: W. H. Freeman and Company, 1965.

Stahl, Eli; Dwyer, Greg; Mauricio, Rodney; Kreitman, Martin & Bergelson, Joy. Dynamics of disease resistance polymorphism at the *Rpm 1* locus of *Arabidopsis. Nature,* 12 August 1999, *400,* 667-671.

Stanley, Steven. *Extinction.* New York: Scientific American Press, 1987.

Steele, Edward; Lindley, Robyn & Blanden, Robert. *Lamarck's Signature.* Reading: Perseus Books, 1998.

Steele, E. J. Lamarck and immunity: Somatic and germline evolution of antibody genes. *Journal of the Royal Society of Western Australia,* 1999, *92,* 437-446.

Stindl, Reinhard. (2014) The telomeric sync model of speciation: species-wide telomere erosion triggers cycles of transposon-mediated genomic rearrangements, which underlie the saltatory appearance of nonadaptive characters. *Naturwissenschaften, 101,* 163-186.

Stone, Richard. Earth's surface may move itself. *Science,* 1 September 1995, *269,* 1214-1215.

Stove, David. The scientific Mafia. In Talbott, Stephen (Ed.) *Velikovsky Reconsidered.* New York: Warner Books, 1976.

Stuller, Jay. Climate is often a matter of inches and a little water. *Smithsonian,* December 1995, 103-

110.
Sudoplatov, Pavel. *Special Tasks.* Toronto: Little, Brown & Company, 1994.

Suvà, Mario; Riggi, Nicolo & Bernstein, Bradley. Epigenetic reprogramming in cancer. *Science,* 2013, *339,* 1567-1570.

Syvanen, Michael. Bacterial insertion sequences. In Kucherlapati, Raju & Smith, Gerald. (Eds) *Genetic Recombination.* Washington D.C.: American Society for Microbiology, 1988.

Terborgh, John. *Diversity and the Tropical Rain Forest.* New York: Scientific American Library, 1992.

Theodorou, V. Susceptibility to stress-induced visceral sensitivity: A bad legacy for next generations. *Neurogastroenterology & Motility,* 2013, *25,* 927-930.

Thompson, D'Arcy. *On Growth and Form.* Mineola: Dover, 1942/1992.

Thomson, Keith. *Living Fossil.* New York: W. W. Norton, 1991.

Tinbergen, N. (1963) On aims and methods of Ethology. *Zeitschrift für Tierpsychologie, 20,* 410-33.

Tindol, Robert. Study links origin of sexual reproduction with high mutation rates. http://news.nationalgeographic.com/news/2001/07/0709_sexorigin.html.

Tregenza, Tom & Butlin, Roger. Speciation without isolation. *Nature,* 22 July 1999, *400,* 311-312.

Trivedi, Bijal. Bugs fighting back in evolutionary war on humans? http://news.nationalgeographic.com/news/2001/0

9/0907_TVbugwar.html.

Tyson, Neil. Vagabonds in space. *Natural History,* July-August 2004, *113,* 16-20.

----------The Perimeter of Ignorance. *Natural History,* November 2005, *114,* 28-34.

Valenkin, Alex. Beyond the Big Bang. *Natural History,* July-August 2006, *115,* 42-47.

Van Decar, John. Crater row. *Nature,* 12 March 1998, *392,* 131.

van Oosterzee, Penny. *When Worlds Collide.* Ithaca: Cornell University Press, 1997.

Vane-Wright, Dick. Butterflies at that awkward age. *Nature,* April 2004, *428,* 478-480.

Verissimo, Adalberto; Chochrane, Mark & Souza, Carlos. National forests in the Amazon. *Science,* 30 August 2002, *297,* 1478.

Vermeij, Geerat J. The Mesozoic marine revolution: evidence from snails, predators and grazers. *Paleobiology,* 1977. *3,* 245-258.

Vernadsky, Vladimir. *The Biosphere.* New York: Copernicus, 1926/1997.

Vogel, Gretchen. The inner lives of sponges. *Science,* 23 May 2008, *320,* 1028-30.

Von Frese, Ralph; Potts, Laramie V.; Wells, Stuart B.; Leftwich, Timothy E.; Kim, Hyung Rae; Kim, Jeong Woo; Golynsky, Alexander V.; Hernandez, Orlando & Gaya-Piqué, Luis R. GRACE gravity evidence for an impact basin in Wilkes Land, Antarctica. *G^3 Geochemistry, Geophysics, Geosystems*, 2009, *10,* 1-14.

Wahlberg, Niklas; Moilanen, Atte & Hanski, Ilka. Predicting the occurrence of endangered species in fragmented landscapes. *Science,* 13 September 1996, *273,* 1536-1538.

Wallace, Alfred R. *Island Life.* Amherst: Prometheus Books, 1881/1998.

---------- *The Malay Archipelago.* Singapore: Periplus, 1890/2000

Wallis, Claudia. The evolution wars. *Time,* 15 August 2005, *166,* 26-35.

Walker, Gabrielle. *Snowball Earth.* New York: Crown Publishers, 2003.

Wakeford, Tom. *Liaisons of Life.* Chichester: John Wiley & Sons, 2001.

Waterland, R. A. & Jirtle, R. A. Transposable elements. *Molecular Cell Biology,* 2003, *23,* 5293-5300.

Watson, J. *Behaviorism.* New York: W. W. Norton, 1924.

Watters, Ethan. DNA is not destiny. *Discover,* October 2006, *27,* 33-37,75.

Weaver, Ian; Cervoni, Nadia; Champagne, Frances; D'Alessio, Ana; Sharma, Shakti; Seckl, Jonathan; Dymov, Sergiy; Szyf, Moshe & Meaney, Michael J. Epigenetic programming by maternal behavior. *Nature Neuroscience, 7,* 847-854.

Weihaupt, John. The Wilkes Land Anomaly: Evidence for a possible hypervelocity impact crater. *Journal of Geophysical Research,* 1976, *81,* 5651–5663.

Weiner, Jonathan. *The Beak of the Finch.* New York: Vintage Books, 1994.

---------- Evolution in action. *Natural History,* November 2005, *114,* 47-51.

Wei, Yong; Pu, Zuyin; Zong, Qiugang; Wan, Weizing;

Ren, Zhipeng; Fraenz, Markus; Dubinin, Eduard; Tian, Feng; Shi, Quanqi; Fu, Suiyan & Hong, Minghua. Oxygen escape from the Earth during geomagnetic reversals: Implications to mass extinction. *Earth and Planetary Science Letters*, 2014, *394,* 94-98.

Weissert, Helmut & Bernasconi, Stefano. An Earth on fire. *Nature,* 11 March 2004, 130- 131.

Weissman, Paul. The Oort Cloud. *Scientific American,* September 1988, *274,* 84-89.

Welch, John. What's wrong with evolutionary biology? *Biol. Philos.,* 2017, *32,* 263-279.

Whiteside, Jessica; Olsen, Paul; Eglinton, Timothy; Brookfield, Michael & Sambrotto, Raymond. Compound-specific carbon isotopes from Earth's largest flood basalt eruptions directly linked to the end-Triassic mass extinction. *PNAS,* 13 April 2010, *107,* 6721-6725.

Wignall, Paul. Large igneous provinces and mass extinctions. *Earth-Science Reviews,* 2001, *53,* 1-33.

----------*The Worst of Times.* Princeton: Princeton University Press, 2015.

---------- Sun, Yadong; Bond, David; Izon, Gareth; Newton, Robert; Védrine, Stéphanie; Widdowson, Mike; Ali, Jason; Ali. Xulong; Jiang, Haisui; Cope, Helen & Bottrell, Simon. Volcanism, mass extinction, and carbon isotope fluctuations in the Middle Permian of China. *Science,* 2009, *324,* 1179-1182.

Williams, Juliet. Scientists believe meteorite his Wisconsin. http://aolsvc.news.aol.com/

news/article.adp?id=20040426122609990001.

Williams, Nigel. Streetcar carries evolution modelers around roadblocks. *Science,* 8 March 1996, *271,* 1365-1366.

Williamson, P. G. Paleontological documentation in Cenozoic mollusks from Turkana Basin. *Nature,* 1981, *293,* 437-443.

Winchester, Simon. *Krakatoa, The Day the World Exploded.* New York: Harper Collins, 2003.

Winn, Ralph. *Psychotherapy in the Soviet Union.* London: Evergreen Books, 1962.

Winters, Jeffrey. Dinosaurs' rise and dominance linked to an earlier asteroid hit. *Discover,* January 2003, *24,* 76.

Wittke, James H.; Weaver, James C.; Bunch, Ted E.; Kennett, James P.; Kennett, Douglas J.; Moore, Andrew M. T.; Hillman, Gordon C.; Tankersley, Kenneth B.; Goodyear, Albert C.; Moore, Christopher R.; Daniel, I. Randolph; Ray, Jack H.; Lopinot, Neal H.; Ferraro, David; Israde-Alcántara, Isabel; Bischoff, James L.; DeCarli, Paul S.; Hermes, Robert E.; Kloosterman, Johan B.; Revay, Zsolt; Howard, George A.; Kimbel, David R.; Kletetschka, Gunther; Nabelek, Ladislav; Lipo, Carl P.; Sakai, Sachiko; West, Allen & Firestone, Richard B. Evidence for deposition of 10 million tonnes of impact spherules across four continents 12,800 y ago. *PNAS,* 4 June 2013. https://www.pnas.org/content/pnas/110/23/E2088.full.pdf

Wynn, Clive. The perils of anthropomorphism. *Nature,* April 8, 2004, 27.
Woit, Peter. *Not Even Wrong.* New York: Basic, 2006.
Wolfsheimer, Gene. *Enjoy the Fighting Fish from Siam.* Harrison: The Pet Library, Ltd. 1975.
Wootton, David. *The Invention of Science.* New York: Harper Perennial, 2016.
Wright, Karen. The day everything died. *Discover*, April 2005, 64-71.
Young, Emma. Rewriting Darwin: The new non-genetic inheritance. *New Scientist,* 2008; *199,* 28-33.
Zahavi, Amotz & Zahavi, Avishag. *The Handicap Principle.* Oxford: Oxford University Press, 1977.
Zamora, Antonio. A model for the geomorphology of the Carolina Bays. *Geomorphology,* 2017, *282,* 209-216.
Zilber-Rosenberg, Ilana & Rosenberg, Eugene. Role of microorganisms in the evolution of animals and plants: the hologenome theory of evolution. *FEMS Microbiology Review,* 2008, *32,* 723-235.
Zimmer, Carl. *Parasite Rex.* New York: Touchstone, 2001 (a).
----------- *Evolution, The Triumph of an Idea.* London, Heinemann, 2001 (b)
----------- Dinosaurs. *Discover,* April 2005, *26,* 32-39.
----------A fin is a limb is a wing. *Natural Geographic*, November 2006, *210,* 111-135.
Zirkle, Conway. The early history of the idea of the inheritance of acquired characters pangenesis. *Trans. Amer. Phil. Soc.,* 1946, *35,* 91-151.

---------*Evolution, Marxian Biology and the Social Scene.* Philadelphia: University of Pennsylvania Press, 1959.

> I ask you to simply open your mind to this possibility. It involves a certain effort.
> ---William Bateson

> Creativity is just connecting things. When you ask creative people how they did something, they feel a little guilty because they didn't really do it, they just saw something. It seemed obvious to them after a while.
> --- Steve Jobs

In politics, the worship of inflexibility or resistance to change is considered a virtue, whereas in science it is almost always an unmistakable sign of pride or shortsightedness. Flexibility is one of the features that best conveys and investigator's honesty. In our view, he who cannot abandon a false concept brands himself as either stupid, dated, or ignorant. Only fools and those who don't read persist in error. Those who insist on being inflexible at all costs seem to declare, with their Olympian disdain for all scientific innovation: "I am worthy, and I know so much that no matter what science discovers, it will not force me to change my views one iota."
---Santiago Ramón y Cajal, *Advice for a Young Investigator*

Ideas that have outlived their day may hobble about the world for years---may even, like Christ, appear after death once or twice to their devotees; but it is hard for them ever again to lead and dominate life. Such ideas never gain complete possession of a man, or they gain possession only of incomplete people.
---Alexander Herzen, *My Past and Thoughts*

Everyone now know that the globe we live on displays almost everywhere the indisputable traces of vast revolutions; the varied products of living nature that embellish its surface are just covering debris that bears witness to the destruction of an earlier nature.
---Georges Cuvier

APPENDIX

On the Law which has Regulated the Introduction of
New Species
Alfred Russel Wallace

Every naturalist who has directed his attention to the subject of the geographical distribution of animals and plants, must have been interested in the singular facts which it presents. Many of these facts are quite different from what would have been anticipated, and have hitherto been considered as highly curious, but quite inexplicable. None of the explanations attempted from the time of Linnaeus are now considered at all satisfactory; none of them have given a cause sufficient to account for the facts known at the time, or comprehensive enough to include all the new facts which have since been, and are daily being added. Of late years, however, a great light has been thrown upon the subject by geological investigations, which have shown that the present state of the earth and of the organisms now inhabiting it, is but the last stage of a long and uninterrupted series of changes which it has undergone, and consequently, that to endeavour to explain and account for its present condition without any reference to those changes (as has frequently been done) must lead to very imperfect and erroneous conclusions.
The facts proved by geology are briefly these:- That during an immense, but unknown period, the surface of the earth has undergone successive changes; land has sunk beneath the ocean, while fresh land has risen up from it; mountain chains have been elevated; islands

have been formed into continents, and continents submerged till they have become islands; and these changes have taken place, not once merely, but perhaps hundreds, perhaps thousands of times:- That all these operations have been more or less continuous, but unequal in their progress, and during the whole series the organic life of the earth has undergone a corresponding alteration. This alteration also has been gradual, but complete; after a certain interval not a single species existing which had lived at the commencement of the period. This complete renewal of the forms of life also appears to have occurred several times:- That from the last of the geological epochs to the present or historical epoch, the change of organic life has been gradual: the first appearance of animals now existing can in many cases be traced, their numbers gradually increasing in the more recent formations, while other species continually die out and disappear, so that the present condition of the organic world is clearly derived by a natural process of gradual extinction and creation of species from that of the latest geological periods. We may therefore safely infer a like gradation and natural sequence from one geological epoch to another.

Now, taking this as a fair statement of the results of geological inquiry, we see that the present geographical distribution of life upon the earth must be the result of all the previous changes, both of the surface of the earth itself and of its inhabitants. Many causes, no doubt, have operated of which we must ever remain in ignorance, and we may, therefore, expect to find many details very difficult of explanation, and in attempting to give one, must allow ourselves to call into our service geological changes which it is highly probable may have occurred, though we have no direct evidence of their individual operation.

The great increase of our knowledge within the last twenty years, both of the present and past history of the organic world, has accumulated a body of facts which should afford a sufficient foundation for a comprehensive law embracing and explaining them all, and giving a direction to new researches. It is about ten years since the idea of such a law suggested itself to the writer of this essay, and he has since taken every opportunity of testing it by all the newly-ascertained facts with which he has become acquainted, or has been able to observe himself. These have all served to convince him of the correctness of his hypothesis. Fully to enter into such a subject would occupy much space, and it is only in consequence of some views having been lately promulgated, he believes, in a wrong direction, that he now ventures to present his ideas to the public, with only such obvious illustrations of the arguments and results as occur to him in a place far removed from all means of reference and exact information.

The following propositions in Organic Geography and Geology give the main facts on which the hypothesis is founded.

Geography

1. Large groups, such as classes and orders, are generally spread over the whole earth, while smaller ones, such as families and genera, are frequently confined to one portion, often to a very limited district.

2. In widely distributed families the genera are often limited in range; in widely distributed genera, well

marked groups of species are peculiar to each geographical district.

3. When a group is confined to one district, and is rich in species, it is almost invariably the case that the most closely allied species are found in the same locality or in closely adjoining localities, and that therefore the natural sequence of the species by affinity is also geographical.

4. In countries of a similar climate, but separated by a wide sea or lofty mountains, the families, genera and species of the one are often represented by closely allied families, genera and species peculiar to the other.

Geology

5. The distribution of the organic world in time is very similar to its present distribution in space.

6. Most of the larger and some small groups extend through several geological periods.

7. In each period, however, there are peculiar groups, found nowhere else, and extending through one or several formations.

8. Species of one genus, or genera of one family occurring in the same geological time, are more closely allied than those separated in time.

9. As generally in geography no species or genus occurs in two very distant localities without being also found in intermediate places, so in geology the life of a species or genus has not been interrupted. In other words, no group or species has come into existence twice.

10. The following law may be deduced from these facts:- Every species has come into existence coincident both in space and time with a pre-existing closely allied species.

This law agrees with, explains and illustrates all the facts connected with the following branches of the subject:- 1st. The system of natural affinities. 2nd. The distribution of animals and plants in space. 3rd. The same in time, including all the phaenomena of representative groups, and those which Professor Forbes supposed to manifest polarity. 4th. The phaenomena of rudimentary organs. We will briefly endeavour to show its bearing upon each of these.

If the law above enunciated be true, it follows that the natural series of affinities will also represent the order in which the several species came into existence, each one having had for its immediate antitype a closely allied species existing at the time of its origin. It is evidently possible that two or three distinct species may have had a common antitype, and that each of these may again have become the antitypes from which other closely allied species were created. The effect of this would be, that so long as each species has had but one new species formed on its model, the line of affinities will be simple, and may be represented by placing the several species in direct succession in a straight line. But if two or more species have been independently formed on the plan of a common antitype, then the series of affinities will be compound, and can only be represented by a forked or many branched line. Now, all attempts at a Natural classification and arrangement of organic beings show, that both these plans have obtained in creation. Sometimes the series of affinities can be well represented for a space by a direct progression from species to species or from group to group, but it is generally found impossible so to continue. There constantly occur two or more modifications of an organ or modifications of two distinct organs, leading us on to two distinct series of species, which at length differ so much from each other as to form distinct genera or families. These are the parallel series or representative groups of naturalists, and they often occur in different countries, or are found fossil in different formations. They are said to have an analogy to each other when they are so far removed from their common antitype as to differ in many important points of structure, while they still preserve a family resemblance. We thus see how difficult it is to determine in every case whether a given relation is an analogy or an affinity, for it is evident that as we go back along the parallel or divergent series, towards the common antitype, the analogy which existed between the two groups becomes an affinity. We are also made aware of the difficulty of arriving at a true classification, even in a small and perfect group;- in the actual state of nature it is almost impossible, the species being so numerous and the modifications of form and structure so varied, arising probably from the immense number of species which have served as antitype for the existing species, and thus produced a complicated branching of the lines of affinity, as intricate as the twigs of a gnarled oak or the vascular system of the human body. Again, if we consider that we have only fragments of this vast system, the stem and main branches being represented by extinct species of which we have no knowledge, while a vast mass of limbs and boughs and minute twigs and scattered leaves is what we have to place in order, and determine the true position each originally occupied with regard to the others, the whole difficulty of the true Natural System of classification becomes apparent to us.

We shall thus find ourselves obliged to reject all those systems of classification which arrange species or groups in circles, as well as those which fix a definite number for the divisions of each group. The latter class have been very generally rejected by naturalists, as contrary to nature, notwithstanding the ability with which they have been advocated; but the circular system of affinities seems to have obtained a deeper hold, many eminent naturalists having to some extent adopted it. We have, however, never been able to find a case in which the circle has been closed by a direct and close affinity. In most cases a palpable analogy has been substituted, in others the affinity is very obscure or altogether doubtful. The complicated branching of the lines of affinities in extensive groups must also afford great facilities for giving a show of probability to any such purely artificial arrangements. Their death-blow was given by the admirable paper of the lamented Mr. Strickland, published in the "Annals of Natural History," in which he so cleverly showed the true synthetical method of discovering the Natural System.

If we now consider the geographical distribution of animals and plants upon the earth, we shall find all the facts beautifully in accordance with, and readily explained by, the present hypothesis. A country having species, genera, and whole families peculiar to it, will be the necessary result of its having been isolated for a long period, sufficient for many series of species to have been created on the type of pre-existing ones, which, as well as many of the earlier-formed species, have become extinct, and thus made the groups appear isolated. If in any case the antitype had an extensive range, two or more groups of species might have been formed, each varying from it in a different manner, and thus producing several representative or analogous groups. The Sylviadae of Europe and the Sylvicolidae of North America, the Heliconidae of South America and the Euploeas of the East, the group of Trogons inhabiting Asia, and that peculiar to South America, are examples that may be accounted for in this manner.

Such phaenomena as are exhibited by the Galápagos Islands, which contain little groups of plants and animals peculiar to themselves, but most nearly allied to those of South America, have not hitherto received any, even a conjectural explanation. The Galápagos are a volcanic group of high antiquity, and have probably never been more closely connected with the continent than they are at present. They must have been first peopled, like other newly-formed islands, by the action of winds and currents, and at a period sufficiently remote to have had the original species die out, and the modified prototypes only remain. In the same way we can account for the separate islands having each their peculiar species, either on the supposition that the same original emigration peopled the whole of the islands with the same species from which differently modified prototypes were created, or that the islands were successively peopled from each other, but that new species have been created in each on the plan of the pre-existing ones. St. Helena is a similar case of a very ancient island having obtained an entirely peculiar, though limited, flora. On the other hand, no example is known of an island which can be proved geologically to be of very recent origin (late in the Tertiary, for instance), and yet possess generic or family groups, or even many species peculiar to itself.

When a range of mountains has attained a great elevation, and has so remained during a long geological period, the species of the two sides at and near their bases will be often very different, representative species of some genera occurring, and even whole genera being peculiar to one side, as is remarkably seen in the case of the Andes and Rocky Mountains. A similar phaenomena occurs when an island has been separated from a continent at a very early period. The shallow sea between the Peninsula of Malacca, Java, Sumatra and Borneo was probably a continent or large island at an early epoch, and may have become submerged as the volcanic ranges of Java and Sumatra were elevated. The organic results we see in the very considerable number of species of animals common to some or all of these countries, while at the same time a number of closely allied representative species exist peculiar to each, showing that a considerable period has elapsed since their separation. The facts of geographical distribution and of geology may thus mutually explain each other in doubtful cases, should the principles here advocated be clearly established.

In all those cases in which an island has been separated from a continent, or raised by volcanic or coralline action from the sea, or in which a mountain-chain has been elevated in a recent geological epoch, the phaenomena of peculiar groups or even of single representative species will not exist. Our own island is an example of this, its separation from the continent being geologically very recent, and we have consequently scarcely a species which is peculiar to it; while the Alpine range, one of the most recent mountain elevations, separates faunas and floras which scarcely differ more than may be due to climate and latitude alone.

The series of facts alluded to in Proposition (3), of closely allied species in rich groups being found geographically near each other, is most striking and important. Mr. Lovell Reeve has well exemplified it in his able and interesting paper on the Distribution of the Bulimi. It is also seen in the Hummingbirds and Toucans, little groups of two or three closely allied species being often found in the same or closely adjoining districts, as we have had the good fortune of personally verifying. Fishes give evidence of a similar kind: each great river has its peculiar genera, and in more extensive genera its groups of closely allied species. But it is the same throughout Nature; every class and order of animals will contribute similar facts. Hitherto no attempt has been made to explain these singular phaenomena, or to show how they have arisen. Why are the genera of Palms and of Orchids in almost every case confined to one hemisphere? Why are the closely allied species of brownbacked Trogons all found in the East, and the green-backed in the West? Why are the Macaws and the Cockatoos similarly restricted? Insects furnish a countless number of analogous examples;- the Goliathi of Africa, the Ornithopterae of the Indian Islands, the Heliconidae of South America, the Danaidae of the East, and in all, the most closely allied species found in geographical proximity. The question forces itself upon every thinking mind,- why are these things so? They could not be as they are had no law regulated their creation and dispersion. The law here enunciated not merely explains, but necessitates the facts we see to exist, while the vast and long-continued geological changes of the earth readily account for the exceptions and apparent discrepancies that here and there occur. The writer's object in putting forward his views in the present imperfect manner is to submit them to the test of other minds, and to be made aware of all the facts supposed to be inconsistent with them. As his hypothesis is one which claims acceptance solely as explaining and connecting facts which exist in nature, he expects facts alone to be brought to disprove it, not a priori arguments against its probability.

The phaenomena of geological distribution are exactly analogous to those of geography. Closely allied species are found associated in the same beds, and the change from species to species appears to have been as gradual in time as in space. Geology, however, furnishes us with positive proof of the extinction and production of species, though it does not inform us how either has taken place. The extinction of species, however, offers but little difficulty, and the modus operandi has been well illustrated by Sir C. Lyell in his admirable "Principles." Geological changes, however gradual, must occasionally have modified external conditions to such an extent as to have rendered the existence of certain species impossible. The extinction would in most cases be effected by a gradual dying-out, but in some instances there might have been a sudden destruction of a species of limited range. To discover how the extinct species have from time to time been replaced by new ones down to the very latest geological period, is the most difficult, and at the same time the most interesting problem in the natural history of the earth. The present inquiry, which seeks to eliminate from known facts a law which has determined, to a certain degree, what species could and did appear at a given epoch, may, it is hoped, be considered as one step in the right direction towards a complete solution of it.

Much discussion has of late years taken place on the question, whether the succession of life upon the globe has been from a lower to a higher degree of organization. The admitted facts seem to show that there has been a general, but not a detailed progression. Mollusca and Radiata existed before Vertebrata, and the progression from Fishes to Reptiles and Mammalia, and also from the lower mammals to the higher, is indisputable. On the other hand, it is said that the Mollusca and Radiata of the very earliest periods were more highly organized than the great mass of those now existing, and that the very first fishes that have been discovered are by no means the lowest organised of the class. Now it is believed the present hypothesis will harmonize with all these facts, and in a great measure serve to explain them; for though it may appear to some readers essentially a theory of progression, it is in reality only one of gradual change. It is, however, by no means difficult to show that a real progression in the scale of organization is perfectly consistent with all the appearances, and even with apparent retrogression, should such occur.

Returning to the analogy of a branching tree, as

the best mode of representing the natural arrangement of species and their successive creation, let us suppose that at an early geological epoch any group (say a class of the Mollusca) has attained to a great richness of species and a high organization. Now let this great branch of allied species, by geological mutations, be completely or partially destroyed. Subsequently a new branch springs from the same trunk, that is to say, new species are successively created, having for their antitypes the same lower organized species which had served as the antitypes for the former group, but which have survived the modified conditions which destroyed it. This new group being subject to these altered conditions, has modifications of structure and organization given to it, and becomes the representative group of the former one in another geological formation. It may, however, happen, that though later in time, the new series of species may never attain to so high a degree of organization as those preceding it, but in its turn become extinct, and give place to yet another modification from the same root, which may be of higher or lower organization, more or less numerous in species, and more or less varied in form and structure than either of those which preceded it. Again, each of these groups may not have become totally extinct, but may have left a few species, the modified prototypes of which have existed in each succeeding period, a faint memorial of their former grandeur and luxuriance. Thus every case of apparent retrogression may be in reality a progress, though an interrupted one: when some monarch of the forest loses a limb, it may be replaced by a feeble and sickly substitute. The foregoing remarks appear to apply to the case of the Mollusca, which, at a very early period, had reached a high organization and a great development of forms and species in the testaceous Cephalopoda. In each succeeding age modified species and genera replaced the former ones which had become extinct, and as we approach the present aera, but few and small representatives of the group remain, while the Gasteropods and Bivalves have acquired an immense preponderance. In the long series of changes the earth has undergone, the process of peopling it with organic beings has been continually going on, and whenever any of the higher groups have become nearly or quite extinct, the lower forms which have better resisted the modified physical conditions have served as the antitypes on which to found the new races. In this manner alone, it is believed, can the representative groups at successive periods, and the rising and fallings in the scale of organization, be in every case explained.

The hypothesis of polarity, recently put forward by Professor Edward Forbes to account for the abundance of generic forms at a very early period and at present, while in the intermediate epochs there is a gradual diminution and impoverishment, till the minimum occurred at the confines of the Palaeozoic and Secondary epochs, appears to us quite unnecessary, as the facts may be readily accounted for on the principles already laid down. Between the Palaeozoic and Neozoic periods of Professor Forbes, there is scarcely a species in common, and the greater part of the genera and families also disappear to be replaced by new ones. It is almost universally admitted that such a change in the organic world must have occupied a vast period of time. Of this interval we have no record; probably because the whole area of the early formations now exposed to our researches was elevated at the end of the Palaeozoic period, and remained so through the interval required for the organic changes which resulted in the fauna and flora of the Secondary period. The records of this interval are buried beneath the ocean which covers three-fourths of the globe. Now it appears highly probable that a long period of quiescence or stability in the physical conditions of a district would be most favourable to the existence of organic life in the greatest abundance, both as regards individuals and also as to variety of species and generic group, just as we now find that the places best adapted to the rapid growth and increase of individuals also contain the greatest profusion of species and the greatest variety of forms,- the tropics in comparison with the temperate and arctic regions. On the other hand, it seems no less probable that a change in the physical conditions of a district, even small in amount if rapid, or even gradual if to a great amount, would be highly unfavourable to the existence of individuals, might cause the extinction of many species, and would probably be equally unfavourable to the creation of new ones. In this too we may find an analogy with the present state of our earth, for it has been shown to be the violent extremes and rapid changes of physical conditions, rather than the actual mean state in the temperate and frigid zones, which renders them less prolific than the tropical regions, as exemplified by the great distance beyond the tropics to which tropical forms penetrate when the climate is equable, and also by the richness in species and forms of tropical mountain regions which principally differ from the temperate zone in the uniformity of their climate. However this may be, it seems a fair assumption that during a period of geological repose the new species which we know to have been created would have appeared, that the creations would then exceed in number the extinctions, and therefore the number of species would increase. In a period of geological activity, on the other hand, it seems probable that the extinctions might exceed the creations, and the number of species consequently diminish. That such effects did take place in connexion with the causes to which we have imputed them, is shown in the case of the Coal formation, the faults and contortions of which show a period of great activity and violent convulsions, and it is in the formation immediately succeeding this that the poverty of forms of life is most apparent. We have then only to suppose a long period of somewhat similar action during the vast unknown interval at the termination of the Palaeozoic period, and then a decreasing violence or rapidity through the Secondary period, to allow for the gradual repopulation of the earth with varied forms, and the whole of the facts are explained. We thus have a clue to the increase of the forms of life during certain periods, and their decrease during others, without recourse to any causes but these we know to have existed, and to effects fairly deducible from them. The precise manner in which the geological changes of the early formations were effected is so extremely obscure, that when we can explain important facts by a retardation at one time and an acceleration at another of a process which we know from its nature and from observation to have been unequal,- a cause so simple may surely be preferred to one so obscure and hypothetical as polarity.

I would also venture to suggest some reasons against the very nature of the theory of Professor Forbes. Our knowledge of the organic world during any geological epoch is necessarily very imperfect. Looking at the vast numbers of species and groups that have been discovered by geologists, this may be doubted; but we should compare their numbers not merely with those that now exist upon the earth, but with a far larger amount. We have no reason for believing that the number of species on the earth at any former period was much less than at present; at all events the aquatic portion, with which geologists have most acquaintance, was probably often as great or greater. Now we know that there have been many complete changes of species; new sets of organisms have many times been introduced in place of old ones which have become extinct, so that the total amount which have existed on the earth from the earliest geological period must have borne about the same proportion to those now living, as the whole human race who have lived and died upon the earth, to the population at the present time. Again, at each epoch, the whole earth was no doubt, as now, more or less the theatre of life, and as the successive generations of each species died, their exuviae and preservable parts would be deposited over every portion of the then existing seas and oceans, which we have reason for supposing to have been more, rather than less, extensive than at present. In order then to understand our possible knowledge of the early world and its inhabitants, we must compare, not the area of the whole field of our geological researches with the earth's surface, but the area of the examined portion of each formation separately with the whole earth. For example, during the Silurian period all the earth was Silurian, and animals were living and dying, and depositing their remains more or less over the whole area of the globe, and they were probably (the species at least) nearly as varied in different latitudes and longitudes as at present. What proportion do the Silurian districts bear to the whole surface of the globe, land and sea (for far more extensive Silurian districts probably exist beneath the ocean than above it), and what portion of the known Silurian districts has been actually examined for fossils? Would the area of rock actually laid open to the eye be the thousandth or the ten-thousandth part of the earth's surface? Ask the same question with regard to the Oolite or the Chalk, or even to particular beds of these when they differ considerably in their fossils, and you may then get some notion of how small a portion of the whole we know.

But yet more important is the probability, nay almost the certainty, that whole formations containing the records of vast geological periods are entirely buried beneath the ocean, and for ever beyond our reach. Most of the gaps in the geological series may thus be filled up, and vast numbers of unknown and unimaginable animals, which might help to elucidate the affinities of the numerous isolated groups which are a perpetual puzzle to the zoologist, may there be buried, until future revolutions may raise them in their turn above the waters, to afford materials for the study of whatever race of intelligent beings may then have succeeded us. These considerations must lead us to the conclusion, that our knowledge of the whole series of the former inhabitants of the earth is necessarily most imperfect and fragmentary,- as much so as our knowledge of the present organic world would be, were we forced to make our collections and observations only in spots equally limited in area and in number with those actually laid open for the collection of fossils. Now, the hypothesis of Professor Forbes is essentially one that assumes to a great extent the completeness of our knowledge of the whole series of organic beings which have existed on the earth. This appears to be a fatal objection to it, independently of all other considerations. It may be said that the same objections exist against every theory on such a subject, but this is not necessarily the case. The hypothesis put forward in this paper depends in no degree upon the completeness of our knowledge of the former condition of the organic world, but takes what facts we have as fragments of a vast whole, and deduces from them something of the nature and proportions of that whole which we can never know in detail. It is founded upon isolated groups of facts, recognizes their isolation, and endeavours to deduce from them the nature of the intervening portions.

Another important series of facts, quite in accordance with, and even necessary deductions from, the law now developed, are those of rudimentary organs. That these really do exist, and in most cases have no special function in the animal economy, is admitted by the first authorities in comparative anatomy. The minute limbs hidden beneath the skin in many of the snake-like lizards, the anal hooks of the boa constrictor, the complete series of jointed finger-bones in the paddle of the Manatus and whale, are a few of the most familiar instances. In botany a similar class of facts has long been recognised. Abortive stamens, rudimentary floral envelopes and undeveloped carpels, are of the most frequent occurrence. To every thoughtful naturalist the question must arise, What are these for? What have they to do with the great laws of creation? Do they not teach us something of the system of Nature? If each species has been created independently, and without any necessary relations with pre-existing species, what do these rudiments, these apparent imperfections mean? There must be a cause for them; they must be the necessary results of some great natural law. Now, if, as it has been endeavoured to be shown, the great law which has regulated the peopling of the earth with animal and vegetable life is, that every change shall be gradual; that no new creature shall be formed widely differing from anything before existing; that in this, as in everything else in Nature, there shall be gradation and harmony,- then these rudimentary organs are necessary, and are an essential part of the system of Nature. Ere the higher Vertebrata were formed, for instance, many steps were required, and many organs had to undergo modifications from the rudimental condition in which only they had as yet existed. We still see remaining an antitypal sketch of a wing adapted for flight in the scaly flapper of the penguin, and limbs first concealed beneath the skin, and then weakly protruding from it, were the necessary gradations before others should be formed fully adapted for locomotion. Many more of these modifications should we behold, and more complete series of them, had we a view of all the forms which have ceased to live.

The great gaps that exist between fishes, reptiles, birds, and mammals would then, no doubt, be softened down by intermediate groups, and the whole organic world would be seen to be an unbroken and harmonious system.

It has now been shown, though most briefly and imperfectly, how the law that "Every species has come into existence coincident both in time and space with a pre-existing closely allied species," connects together and renders intelligible a vast number of independent and hitherto unexplained facts. The natural system of arrangement of organic beings, their geographical distribution, their geological sequence, the phaenomena of representative and substituted groups in all their modifications, and the most singular peculiarities of anatomical structure, are all explained and illustrated by it, in perfect accordance with the vast mass of facts which the researches of modern naturalists have brought together, and, it is believed, not materially opposed to any of them. It also claims a superiority over previous hypotheses, on the ground that it not merely explains, but necessitates what exists. Granted the law, and many of the most important facts in Nature could not have been otherwise, but are almost as necessary deductions from it, as are the elliptic orbits of the planets from the law of gravitation.

Sarawak, Borneo, Feb 1855

www.ingramcontent.com/pod-product-compliance
Lightning Source LLC
Chambersburg PA
CBHW060821170526
45158CB00001B/50